CONSUMER SENSORY TESTING FOR PRODUCT DEVELOPMENT

CONSUMER
SENSORY TESTING
·····FOR·····
PRODUCT
DEVELOPMENT

Anna V. A. Resurreccion
The University of Georgia

A Chapman & Hall Food Science Book

An Aspen Publication®
Aspen Publishers, Inc.
Gaithersburg, Maryland
1998

The author has made every effort to ensure the accuracy of the information herein. However, appropriate information sources should be consulted, especially for new or unfamiliar procedures. It is the responsibility of every practitioner to evaluate the appropriateness of a particular opinion in the context of actual clinical situations and with due consideration to new developments. The author, editors, and the publisher cannot be held responsible for any typographical or other errors found in this book.

Library of Congress Cataloging-in-Publication Data

Resurreccion, Anna V.A.
Consumer sensory testing for product development/Anna V.A.
Resurreccion.
p. cm.
Includes bibliographical references and index.
ISBN 0-8342-1209-9 (alk. paper)
1. Food—Sensory evaluation. 2. Commercial products—Testing.
I. Title.
TX546.R47 1998
664'.07—dc21
97-48866
CIP

Orders: (800) 638-8437
Customer Service: (800) 234-1660

About Aspen Publishers • For more than 35 years, Aspen has been a leading professional publisher in a variety of disciplines. Aspen's vast information resources are available in both print and electronic formats. We are committed to providing the highest quality information available in the most appropriate format for our customers. Visit Aspen's Internet site for more information resources, directories, articles, and a searchable version of Aspen's full catalog, including the most recent publications: **http://www.aspenpub.com**
Aspen Publishers, Inc. • The hallmark of quality in publishing
Member of the worldwide Wolters Kluwer group.

Editorial Services: Ruth Bloom
Library of Congress Catalog Card Number: 97-48866
ISBN: 0-8342-1209-9

Printed in the United States of America

1 2 3 4 5

To
Rey

Contents

Preface

Consumer affective tests are necessary in product evaluation for product development guidance, product improvement and optimization, and maintenance. The testing methods employ untrained individuals and larger sample sizes than required in sensory analytical test methods. Consumer affective tests appear to be relatively simple to plan and conduct. Unfortunately, the seemingly easy task is often invalidated by improper data-collection methodology, which invariably leads to faulty interpretation of results. The validity and reliability of the consumer testing methodologies are extremely important if test results will be used as a basis for major business decisions.

The objective of this book is to provide information on recommended methods and procedures in the planning and implementation of consumer tests, and data collection and analysis of consumer test results. This is primarily a reference book for the beginning sensory scientist and food industry personnel, university and government researchers involved in product quality evaluations, consumer affective tests, and product research and development. It is intended to assist the sensory practitioner in conducting a consumer affective test project and an in-house sensory team to carry out the tests; if the sensory tests are to be conducted by another agency, this book should assist the project leader in planning and monitoring the test.

A chapter on consumer affective testing is a part of almost every existing sensory evaluation text. However, the intent of most of those chapters is to focus on general principles but do not provide comprehensive information on the procedures for the conduct of various affective test methods. Individuals with limited experience in consumer affective testing are faced with a considerable search of the literature before they can implement a valid consumer test with confidence. Consumer affective tests are the subject of a fifty-three-page *not available* monograph on "Consumer Sensory Evaluation," published in 1979 by the American Society for Testing and Materials (ASTM), Committee E-18. The monograph covers many important considerations in conducting a consumer test; however, space constraints impose limits on a more comprehensive coverage. Experienced researchers in industrial or academic settings are likewise in need of a reference that combines the basic principles and discusses procedures that are unique to certain target populations such as very young children. The demand for information on consumer testing became more evident as requests for lectures on the topic increased. This book originated from a series of lectures given in a course held yearly at the University of Georgia. Its scope increased with the development of week-long intensive training courses on "consumer testing for new product development" held in the Philippines, Malaysia, and Thailand, attended by over 125 participants to date. The intent of this book is to fill the need for a simple but comprehensive guide to consumer sensory testing for product development.

The organization of the chapters is designed to be straightforward. Chapter 1 is an introduction to consumer testing in product development, maintenance, improvement, and optimization. Chapter 2 discusses affective test methods, data collection, and questionnaire design. Chapter 3 is on test procedures and the planning, design, and implementation of consumer affective tests. Chapter 4, on the consumer panel, discusses sample size, sampling and demographic characteristics on the panel, sources of bias when using an employee panel, recruitment, and maintenance of a consumer database. In Chapter 5, qualitative consumer research methods are described. Quantitative methods are described by test location in Chapter 6—the sensory laboratory test, Chapter 7—central location tests, Chapter 8—mobile laboratory tests, Chapter 9—home-use tests, and Chapter 10—simulated supermarket-setting studies. The advantages and disadvantages of each test are discussed, as are the role of the project leader, the test procedure, the panel, test facility and environment, and special considerations related to the type of test. Chapter 11 is on affective testing with children. The advantages of using age-appropriate scale lengths in testing with children, effect of decorations used in the children's testing area, use of child-sized furniture, and sources of bias are discussed. Hypothesis testing and statistical analysis methods are the subject of Chapter 12, where both univariate and multivariate techniques are described. Finally, Chapter 13 contains a discussion of the methods used in the quantification of quality and relationships between consumer affective test results, and descriptive sensory analysis and physico-chemical measurements. Appendices (A–D) contain various checklists to be used in the planning and implementation of consumer tests. Appendix E has a number of statistical charts.

New approaches to be covered in this book that have not received much attention in publications on consumer testing are (1) methodologies on testing of young children, (2) simulated supermarket-setting tests, and (3) the use of a mobile laboratory in consumer tests. These topics have been the subject of the author's research activities and publications during the past ten years.

With regard to terminology, although an attempt was made to use the term panelist consistently, the terms *subject*, *judge*, *respondent*, and *assessor* may have been used interchangeably. Likewise, the terms *he*, *she*, and *they* were used interchangeably with *project leader*, *sensory scientist*, *sensory practitioner*, *sensory professional*, or *sensory analyst*.

I gratefully acknowledge the assistance and input of many colleagues and students in providing examples to illustrate many of the techniques described. I thank Dr. Tommy Nakayama, my former department head, for his vision regarding a consumer and sensory research program at the University of Georgia; for his guidance and the advice to develop my lectures into material for a book. A very special statement of gratitude goes to Ms. Jye-Yin Liao for providing extraordinary assistance and dedication in the coordination of the technical and secretarial support, and the enormous details required in the preparation of the manuscript, and for preparing the table of contents and figures, and managing the literature databases and reference lists. This book would not have been possible without her able assistance. I thank my mother, Virginía F. Agbayani, for her encouragement and prayers, and my husband, to whom this book is dedicated, for his support during the entire process of writing this book and throughout my career.

1

Introduction

Consumer Testing

Consumer testing is one of the most important activities in product development. The primary purpose of consumer affective tests is to assess the personal response by current and potential customers of a product, product ideas, or specific product characteristics. Consumer evaluation concerns itself with testing certain products using untrained people who are or will become the ultimate users of the product. These products are evaluated on the basis of appearance, taste, smell, touch, and hearing (ASTM, 1979). Validity and reliability of the consumer testing methodologies are extremely important.

Consumer testing is necessary throughout the various stages in the product cycle (ASTM, 1979). These stages include the development of the product itself, product maintenance, product improvement and optimization, and assessment of market potential.

Sensory acceptance tests are conducted during product development for product development guidance, to screen products, to identify those products that are significantly disliked and those that match or exceed a specified target product for acceptance. Sensory tests proceed with the selection of a product or products that will continue on to large-scale testing. These sensory tests are not primarily designed to determine market demand or to be used for market segmentation or related demographic tabulations. Sensory acceptance tests indicate the acceptance of a product without the package, label, price, and so on. The big difference between consumer sensory and market research testing is that the sensory test is generally conducted with coded, unbranded products, whereas market research is most frequently done with branded products (van Trijp and Schifferstein, 1995). One should not expect blind-labeled sensory tests to give the results obtained from market research tests. Each has its purpose in product testing. The implicit goal behind any and all sensory evaluation efforts in the food industry is to enhance quality to improve appearance, flavor, and texture as perceived by consumers in order to influence their food choices (translated into purchases) at the point of sale.

Product Development

Most products created today are designed to satisfy the needs of consumers (ASTM, 1979). Each year, hundreds of food products are introduced and at least 90% quickly disappear, not be seen again for at least another few years (Stone, 1988). Product concepts may arise from a variety of sources, including the marketing and research departments, management input, consumer research, and idea generation. One of the recommended strategies to assess a concept early in the development stage is by using a focus group.

In a focus group, consumer responses are used to qualitatively assess the concept or to identify critical attributes of a product or concept. Beyond the concept development, initial prototype products are developed that possess the critical attributes that are valued by the consumers and lack the attributes that consumers consider unacceptable. At this stage, consumer sensory tests are a useful tool for product guidance. The objective of the research is to assess performance potential and for product guidance for further development with the goal of maximizing acceptance.

At the end of the product development stage, the alternate product prototypes have usually been narrowed down to a manageable number of final prototypes. Analytical tests have been performed to assist in the decision on which products should be tested by consumers. Acceptance tests give an estimate of product acceptance in different areas around the country. The testing methods employ untrained individuals and larger sample sizes than required in sensory analytical test methods. Test participants may be composed of 100–500 consumers in three or four cities throughout the country. Although acceptance tests are an essential part of the decision-making process in new product development, a successful acceptance test does not guarantee success of a product in the marketplace. Sensory acceptance tests are not a substitute for large-scale market tests.

If prior testing has shown adequate acceptance for a product, the decision is made on whether the product will proceed to the test market stage. At this point, execution of test market research is the function of the market research department. Although test market research is beyond the scope of this book, the sensory evaluation, market research, and management teams need to be working in close collaboration to ensure that the proper tests are being conducted.

Product developers will need to scale up production using large-scale manufacturing equipment. Formulation or process changes may require sensory tests to ensure that these changes do not compromise consumer acceptability. In addition, it is important to establish the length of time that a product will remain stable. This type of testing is needed to establish open dating (sell-by or use-by dates) parameters.

Product Maintenance

Consumer sensory evaluation is essential in monitoring product acceptance to establish and maintain position within a market. Consumer affective tests will help to assess changes in product acceptance with time, changes in raw materials, or processes. Home-use tests are often used to monitor the relative acceptance or preference for a food product. Sensory tests may be used to explain changes in market share due to competition (ASTM, 1979). If market share is eroding, consumer sensory tests may provide direction for changes that need to be implemented to increase consumer acceptance.

Product Improvement

After a period of rapidly increasing sales and profits, it is common to observe a leveling of sales increase (ASTM, 1979). Food processors may improve the product by adding new flavors, or introduce line extensions of the basic product. These should be tested as one would a new product. Laboratory tests, central location tests, and home-use tests are used to assess acceptability of the new products against the original and competing products.

Product Optimization

Because of profit decline that accompanies the maturity of a product, cost reductions that do not compromise the acceptance of the product are vital. Changes in raw materials in the formulation and processing should not decrease acceptance of the product. Consumer tests are used to measure any effect of these changes in acceptance of the product.

General Requirements for Sensory Testing

Consumer testing, as in all types of sensory testing, requires special controls and the use of standard practices. Practitioners must realize that when controls are not employed, the validity of the test is compromised. When sensory testing is carried out in facilities where control is difficult, project personnel need to try eliminate sources of bias in responses. The standard practices regarding facilities are mentioned in this chapter. Special problems concerning consumer tests in different locations are mentioned in Chapters 5–10. With the exception of the consumer test that is conducted in the laboratory, less control can be exerted when testing in different locations. The project leader and sensory team working under such conditions must exercise judgment to maintain a valid test. Other standard practices have to do with the selection of samples, their preparation, and presentation, and are discussed in Chapter 3. A discussion on the panel is found in Chapter 4.

Facilities

Location

The test facility should be located as close to potential panelists as possible. The location will determine the ease in obtaining respondents who are typical consumers of the food products to be tested. Inconveniently located facilities will discourage potential panelists from participating in tests. A ground-floor location with entrances next to adequate parking areas is desired. Panelists should not have to pass through the food preparation or office areas. On the other hand, the reception room and booth areas should preferably occupy areas that are distant from high-traffic areas to minimize noise and confusion, but this would sacrifice accessibility of the test location to panelists. A solution would be to locate the test areas for increased accessibility to panelists but employ special procedures to control noise in the booth area panel by sound-proofing.

Layout

In most sensory facilities, the area consists of the booth area, a discussion area, food preparation facilities, and a waiting room for panelists. The design of the layout should have three objectives: (1) efficient physical operations, (2) avoidance of distraction of panelists due to laboratory equipment and personnel, and (3) minimization of distraction among respondents.

Evaluation Area.   The booth area and discussion room should be separated adequately from the kitchen area to prevent the migration of odors from cooking or from highly flavored substances. Partitioned booths are desirable to minimize distraction from other panel members, but these should not leave the panelists with a feeling of isolation. The aisle behind the booths should allow the panelists to comfortably slide in and out without disturbing the other panelists. In the United States, guidelines for aisle widths, counter heights, and seating configurations by the American Disabilities Act of 1990 should be followed to allow panelists with disabilities to participate in the tests. When partitioned booths are not available, temporary booths (Lawless and Heymann, 1977) may be used to minimize distraction between panelists. If temporary booths are not feasible to use, participants should be positioned so that they do not face each other. The furnishings should be a neutral color. When planning for a booth area, the practitioner should attempt to have the maximum possible number of booths, as space will allow. The countertop heights may be desk-height or counter-height. Both configurations have been used. Desk-height counters allow comfortable seating and working for panelists but, depending on the design of the sample pass-through hatch, may require the server to bend when serving the samples. If the pass-through hatch opening is at counter-height, the server will not have to bend when serving samples but only slide these through. A counter-height booth counter will facilitate serving, but panelists will need to be provided with stools, which could be uncomfortable to the panelists due to the stool height. Serving hatches of the sliding door or breadbox style allow the panelists to see through the open space through the counter. On the other hand, these hatches do not take up counter space.

The booth area needs to be equipped with the electrical and network connections for computerized data collection. Although sinks were installed in many sensory booth areas for many years, the use of sinks is no longer encouraged. Sinks are now thought of as a major source of odor contamination. Disposable personal spittoons provide a sanitary alternative.

The use of signal lights to inform servers that the panelists are in the booth area, to signal the need for a sample, or to request for help minimizes panelist interaction with the servers. When the signal light system is not available, a colored card can be slipped under the hatch door to gain attention from sample servers.

Waiting Room. The waiting room should be located away from the booth area to prevent waiting panelists from distracting those in the booth area. The waiting room will be used for social interaction, payment of incentives, or other activities that need to take place prior to the test. It must be comfortable and have adequate lighting. Light reading material such as magazines may be placed in the room for panelists to use. The waiting room is sometimes used as the orientation or the briefing room.

Food-Preparation Areas. These will be designed on the basis of products to be evaluated. There are standard pieces of equipment that should go in every kitchen. Ample cabinet, refrigerated and frozen storage should be designed.

Environmental Control

Odor

The test areas must be as free from odors as possible. A slight positive pressure in the evaluation areas will reduce the migration of odors from the food preparation and other areas. Air from the sample preparation areas should be vented through to the outside of the building. Air entering the evaluation room should preferably pass through activated charcoal filters. All materials and equipment in the room should be odor-free or have a low odor level.

Lighting

Adequate illumination is important in the testing areas. Special light effects such as colored bulbs or sodium lamps may be desired, in some instances, to hide differences in color.

General Comfort

Controlled temperature and humidity will result in comfortable surroundings that encourage concentration of panelists during a test. Furnishings, countertop heights, and computer placement should be ergonomically designed.

Computerization of the Sensory Laboratory

Typically, data from sensory evaluation experiments are collected from panelists' responses on paper ballots. In recent years, a number of sensory laboratories have replaced the traditional method of utilizing paper ballots with computerized, direct data collection. It is generally accepted that computerization of sensory evaluation significantly reduces cost and time required to prepare, collect, and analyze sensory data (Winn, 1988) as well as increases accuracy over data manually entered from paper ballots. Federico (1991) observed that many economic and technological benefits can be gained from computer use. With decreasing computer costs and increasing user knowledge, these benefits are especially achievable. Pecore (1984) stated that use of computers in consumer testing is valuable because the methods require the largest numbers of respondents and often the largest volume of data. Thus, "the ability to store and process a large amount of information makes computerization of the sensory organization a necessity" (Savoca, 1984).

Although computers are becoming more common in business as well as daily life, the sensory scientist should be aware of the implications of making computers an integral part of consumer test data acquisition. When a sensory laboratory relies on computers for data acquisition, Savoca (1984) states that the sensory analyst places greater emphasis on each panelist's ability to assist in the data-collection phase. Panelists have the responsibility of entering their own responses and correcting their own entry errors.

The literature does not agree, according to Meier and Lambert (1991), about such basic issues as the relationship between computer aversion and variables such as age, gender, cognitive abilities, and computer experience. In a study to determine correlates between these variables and three computer aversion scales, the authors determined that students with more computer skills experienced less computer aversion. "Small gender and age differences also appeared, indicating younger and female students were more likely to experience discomfort with computer use" (Meier and Lambert, 1991). In this study, cognitive ability measurements indicated that more intelligent students reported less computer aversion.

Johnson and Johnson (1981) performed psychological studies on considerations such as panelist instructions, panelist error in entering their responses, and availability of simple error-correction procedures that could be explored in relation to the development of computerized testing stations. The authors reported that instructions to panelists should initially be detailed and become progressively less detailed as the users become more familiar with the computer program. Computer system engineers should also be aware that errors in data entry will be made, and a simple correction procedure should be developed. Synodinos et al. (1994) found the effect of the method of using computer-administered versus paper-and-pencil surveys was small when the panelists for the study were randomly selected but not when panelists were self-selected. In recent studies conducted by Plemmons (1997), computer ballot scores were equivalent to paper ballot scores in a consumer sensory evaluation test for foods representing the hedonic scale continuum of dislike extremely ($= 1$) to like extremely ($= 9$). For disliked foods, panelist hedonic scores were equivalent among panelists with computer experience or without. When using a computer for data collection, panelists' food attitude scores were lower than hedonic food scores from a disliked food, equivalent in neutral foods, and higher in liked foods. She concluded that computerized ballots produce results equivalent to paper ballots in a consumer affective test. In addition, because the computer is a considerable time saver in consumer testing, with direct data input using computers to determine differences between treatments or in continuing studies, she recommends their use in tests involving consumers.

References

ASTM, Committee E-18. 1979. *ASTM Manual on Consumer Sensory Evaluation,* ASTM Special Technical Publication 682, E. E. Schaefer, ed. American Society for Testing and Materials, Philadelphia, PA, pp. 28–30.

Federico, P. A. 1991. Measuring recognition performance using computer-based and paper-based methods. Behav. Res. Meth. Instru. 23:341–347.

Johnson, J. H., and Johnson, K. N. 1981. Psychological considerations related to the development of computerized testing stations. *Behav. Res. Meth. Instru.* 13:421–424.

Lawless, H. T., and Heymann, H. 1997. *Sensory Evaluation of Food: Principles and Practices.* Chapman and Hall, New York.

Meier, S. T., and Lambert, M. E. 1991. Psychometric properties and correlates of three computer aversion scales. *Behav. Res. Meth. Instru.* 23:9–15.

Pecore, S. D. 1984. Computer-assisted consumer testing. *Food Technol.* 38(9):78–80.

Plemmons, L. P. 1997. Sensory evaluation methods to improve validity, reliability, and interpretation of panelist responses. M.S. thesis, University of Georgia, Athens, GA.

Savoca, M. R. 1984. Computer applications in descriptive testing. *Food Technol.* 38(9):74–77.

Stone, H. 1988. Using sensory resources to identify successful products. In *Food Acceptability.* D. M. H. Thomson, ed., Elsevier, London. pp. 283–296.

Synodinos, N. E., Papacostas, C. S., and Okimoto, G. M. 1994. Computer-administered versus paper-and-pencil surveys and the effect of sample selection. *Behav. Res. Meth. Instru.* 26:395–401.

van Trijp, H. C. M., and Schifferstein, H. J. N. 1995. Sensory analysis in marketing practice: Comparison and integration. *Journal of Sensory Studies* 9:205–216.

Winn, R. L. 1988. Touch screen system for sensory evaluation. *Food Technol.* 42(11):68–70.

2

Sensory Test Methods

Introduction

Acceptance testing is a valuable and necessary component of every sensory evaluation program. During product evaluation, acceptance testing usually but not always follows discrimination and descriptive tests, which have been used to reduce the large array of samples to a limited number, and precedes larger scale testing conducted outside of research and development by other departments such as the marketing research department (Stone and Sidel, 1993).

It is important to emphasize that the sensory acceptance test is neither a substitute for the large-scale consumer tests, nor is it a competitive alternate. In other words, the sensory test should not be used to take the place of large-scale consumer tests when the latter are needed; one type of test will not replace the other (Stone and Sidel, 1993).

The consumer acceptance test is a small panel test, usually involving only 50–100 panelists (IFT/SED, 1981), whereas the large-scale market test usually requires at least 100 or more panelists from each of several strategically selected geographic locations. Furthermore, the emphasis of sensory evaluation is on the product and on determining the best product for the target market. Consumer acceptance tests determine overall preference or liking of a product, or of a product's sensory properties such as appearance, including color, flavor, and texture (Meilgaard et al., 1991). Market research usually focuses on consumer populations and identifying the consumers to whom the product will appeal, and developing the means to reach those consumers (Stone and Sidel, 1993). Therefore, sensory evaluation and market research are two separate but related activities that complement one another and rely on different testing procedures.

Acceptance tests are conducted to (1) determine overall liking or preference for a product or products by a sample of consumers who represent the population for whom the product is intended, (2) measure liking or preference for a product's sensory properties, including appearance, flavor and texture, and (3) quantify consumer responses by relating these to descriptive analysis results or physical and chemical measurements.

Acceptance and Preference Tests

There are two approaches to consumer sensory tests. These are the measurement of preference and the measurement of acceptance (Jellinek, 1964).

Acceptance Tests

Consumer acceptance of a food may be defined as (1) an experience, or feature of experience, characterized by a positive attitude toward the food; and/or (2) actual utilization (such as purchase or eating) of food by consumers. Acceptance may be measured by preference or liking of a specific food item (Amerine et al., 1965). The measurement of acceptability is inferred from scale ratings. Consumer acceptance tests measure the acceptability or liking for a food. Acceptance measurements can be made on single products and do not require comparison to another product. The questions asked during these tests are "How much do you like the product?" (Stone and Sidel, 1993) or "How acceptable is the product?" (Stone and Sidel, 1993; Meilgaard et al., 1991).

The consumer acceptance test gives an estimate of product acceptance based on the product's sensory properties. This measure of acceptance does not guarantee success in the marketplace since packaging, price, and advertising are among other factors that can have an influence. However, results of the test provide an indication of product acceptance without the effect of the other factors, which can enhance its acceptance.

Preference Tests

Preference tests refer to all affective tests based on a measurement of preference, or a measurement from which relative preference may be determined (IFT/SED, 1981). Preference may be defined as (1) an expression of higher degree of liking; (2) choice of one object over others; and (3) psychological continuum of affectivity (pleasantness/unpleasantness) upon which such choices are based (Amerine et al., 1965). It may include the choice of one sample over another, a ranked order of liking, or an expression of opinion on a hedonic (like/dislike) scale.

Preference tests measure the appeal of one food or food product over another (Stone and Sidel, 1993). The panelist has a choice, and one product is to be chosen over another. The questions asked are "Which sample do you prefer?" or "Which sample do you like better?" (Meilgaard et al., 1991). Preference tests are useful when one product is compared directly against another, such as in product improvement or against competing products. Scaling methods allow us to measure directly degree of liking and to compute preferences from these data.

Preference can be measured directly or indirectly. The measurement of preference from paired comparison or ranking tests is direct, whereas preference from hedonic ratings is implied. It is achieved by determining which product is rated significantly higher (more liked) than another product in a multi-product test, or which product is rated higher than another by significantly more people. There is an obvious and direct relationship between measuring product liking/acceptance and preference (Stone and Sidel, 1993).

Preference methods can be used to determine differences in preference, but not differences per se; discriminative tests should be used for this purpose. In tests to determine whether a product that is the company's gold standard or target product is preferred over products from the different plants or different suppliers, a preference test is appropriate. However, if the objective of the test is to determine whether the gold standard or target product is different from the product from different plants or suppliers, the test should be used as a discriminative test rather than a preference test. Difference tests have been used to detect perceptible differences between two samples prior to conducting a consumer preference test. This is based on the assumption that if a difference does not exist, subsequent preference testing is unnecessary (IFT/SED, 1981).

Methods Used in Acceptance and Preference Tests

The three types of tests used in consumer acceptance testing are (1) the paired preference, (2) ranking, and (3) rating tests. However, two of the most frequently used tests to measure consumer preference and acceptance are the paired preference and the hedonic scale, respectively. Other methods are described in the literature, but many of them are either modifications of these two methods or are types of quality scales such as those that rate "excellent" to "poor" or "palatable" to "unpalatable."

Paired Preference Tests

In this test, two samples are usually presented simultaneously and rarely, sequentially, if samples have a characteristic that would prevent their being presented simultaneously, and the panelists are asked to indicate which of the two products is preferred. This test is easily carried out and works well even when the consumer panelists have minimal reading or comprehension skills. The test instructions may or may not force the panelist to make a decision. If a forced choice is imposed, the panelist may not give a "no preference" response (ASTM, 1996) and must indicate a preference for one sample over another. The method may also be used for multiple paired preferences within a sample series, such as a standard product versus each of several experimental products.

PAIRED PREFERENCE TEST

Name _____   Session Code _____   Date _____

Please rinse your mouth with water before starting.

In front of you are two samples. Beginning with the sample on the left, taste each one and **circle the sample that you prefer.** You must choose one. You may retaste as often as you need to.

           362   **179**

Thank you

Figure 2.1   An example of a ballot to be used in a paired-preference test. This is a forced choice test. The no-preference response is not an option.

One or more sample pairs of products may be tested per panel session; the upper limit for the number of sample pairs that can be tested is determined by physiological and psychological constraints.

The paired preference test is simple and relatively easy to organize and conduct. The two coded products are served together to naive panelists, and each panelist is asked to indicate which sample is preferred after sampling both products. A sample scoresheet is shown in Figure 2.1. It is advisable to include the "no preference" and "dislike both equally" choices among the responses, and to use an adequate sample size when doing so. When using the "no preference" option there are 3 choices for handling the data: These are listed in Table 2.1. Samples are always identified by three-digit random numbers, and the serving order is always balanced, with the product A served first as many times as product B is served first.

Table 2.1   Handling the "No-Preference" Option in Paired-Preference Tests

---

1. Ignore the "no-preference" response.
2. Split the "no-preference" judgments equally.
3. Divide the "no-preference" responses proportionally according to the preference ratio.

Advantages

The advantage of the paired preference tests is that the test is easy to organize and to implement. Only two orders of presentation, A-B and B-A, are possible. However, the order of presentation must be balanced across panelists or position bias can likely occur. One panelist usually evaluates only one pair of products in a test without a replication. If the "no-preference" choice is used, the number of consumers answering a preference for either sample will be smaller; therefore, this option should be used when 100 or more consumers are participating in the test. If the "dislike both equally" choice is used, the analysis should be performed only on the preference responses. It has been suggested (Stone and Sidel, 1993) that if at least 5% of participants select this category, one needs seriously to question the appropriateness of the products for the tests or the sample of consumer panelists used for testing the product.

Disadvantages

From a sensory evaluation standpoint, the paired preference test provides less information than a rating test, because it gives no direct measure of the magnitude of preference. The paired preference test is less efficient than a rating test, because only one response per product pair is obtained, as opposed to one response per product when a rating test is used. If the "no preference" category is not used, it is virtually impossible to determine if both products were disliked. Because preference testing asks the respondent to evaluate the product on a total basis, the testing may be especially susceptible to unintentional biases that can arise from slight differences in placement on the serving tray, serving temperatures, sample volume, and variability of the sample from the manufacturing process, among others (ASTM, 1996).

In a multiproduct test, there will be considerable interaction because of flavor carryover from one product to the next. In addition, memory can be a confounding variable if a sequential presentation is selected for the paired preference test.

Analysis

In the paired preference test, the probability of selection of a specific product is one of two. The null hypothesis states that the consumers will pick each product an equal number of times. The probability of the null hypothesis is $P = 0.5$. If the underlying population does not have a preference for one sample over the other, the probability of choosing product A is equal to that of choosing product B. This may be written as

$$H_0: = p(A) = p(B) = \frac{1}{2}$$

The alternative hypothesis for the paired preference test is that if the underlying population has a preference for one product over the other, the preferred product will be selected more often than the other product. This may be written as

$$H_a: p(A) \neq p(B)$$

The paired preference test is usually a two-tailed test, because we do not have prior knowledge regarding which products will be preferred by the consumer population. In instances where there is interest in only one product, and the hypothesis assumes that one of the products will be preferred (Stone and Sidel, 1993) the test will be based on a one-tailed test. Analysis is accomplished using any of the following statistical methods: binomial, chi-square, or normal distributions or use of tables.

Binomial Probabilities in Paired Preference Tests

A paired preference is conducted to determine whether consumers prefer the freshly squeezed product to an experimental lemonade prepared from a base mixture. If the base mixture is used, it will minimize labor costs and reduce variability in the product. In this test, 60 of 100 consumers preferred the freshly squeezed lemonade formulation. What is the p-value for this test?

The null hypothesis, H_0 is that the population proportion preferring the test item is $p = \frac{1}{2}$. The alternative hypothesis $H_a = p(A) \neq p(B)$.

The Z score is given by the formula

$$Z = \frac{(\text{Proportion observed} - \text{Proportion expected}) - \text{Continuity correction}}{(\text{Standard error of the proportion})}$$

where

Proportion observed = X/N; and X = number of consumers preferring the test product, N = total consumers in panel.

Proportion expected = Chance level = $\frac{1}{2}$ for the paired preference test

Continuity correction = $(\frac{1}{2} \times N)$; standard error of the proportion = $(pq/N)^{\frac{1}{2}}$

p = chance proportion ($= \frac{1}{2}$), and $q = (1 - p)$

If the calculated Z score is > 1.96, we have an observation that would occur less than 1 chance in 20 (5% level of significance) under a true null hypothesis. In this example, if we have 60% preferring the freshly squeezed lemonade,

$$Z = [(60/100) - (\tfrac{1}{2})] - [(1/200)]/[(0.5)/100]^{\frac{1}{2}}$$
$$= 0.095/0.05$$
$$= 1.90$$

Z is not > 1.96, therefore the p-value is less than .05, so there is no evidence to reject the null hypothesis, H_0, and we conclude that there is no perceptible difference in preference between the two products.

Binomial Distribution and Tables

The binomial distribution allows one to determine whether the result of the paired preference test was due to chance or whether the panelists actually preferred one sample to the other. The formula

$$p = [(\tfrac{1}{2})^N \times N!]/[(N - X)! \; X!]$$

where

$\quad$ N = the total number of judgments

$\quad$ X = the total number of preference judgments for the product that was preferred

$\quad$ p = the probability of making the number of preference choices for the preferred sample.

$\quad$ $N!$ or N factorial is calculated as

$$N! = (N) \times (N-1) \times (N-2) \times (N-3). \ldots \ldots \ldots .(2) \times (1)$$

The following worked example uses small numbers to illustrate the calculations that must be made. For example, the exact probability of 4 to 6 panelists preferring the freshly squeezed lemonade to the lemonade from a base mixture is

$$p = (\tfrac{1}{2})^N \times (N!)/[(N - X)! \; (X!)]$$
$$= [(\tfrac{1}{2})^6 \times (6 \times 5 \times 4 \times 3 \times 2 \times 1) \, / \, (2 \times 1)(4 \times 3 \times 2 \times 1)]$$
$$= [.0156 \times 720/(2)(24)]$$
$$= [11.23/48]$$
$$= .234$$

This is the probability of just one outcome, and two other factors need to be taken into account. These are that the probability of the terms that are farther out in the tail of the distribution must be added to the probability (Lawless and Heymann, 1997) and that because we have not made a prediction regarding which sample will be preferred, the test is two-tailed, and the total probability should be doubled.

To calculate the exact probability in the tail of the distribution, each term is calculated by

$$p = [(½)^X (½)^{N-X}] \times [(N!)/N - X)! (X!)]$$

where

X = the number preferring
N = the total number of panelists
p = the chance level of ½
$q = (1 - p)$

This must be calculated for 6/6 preferring a sample to 4/6 preferring a sample to get the entire tail.

Therefore,

$$\text{for } 6/6: p = [(½)^6(½)^{6-6}] \times [(6!)/(6-6)! (6!)]$$
$$= (½)^6$$
$$\text{for } 5/6: p = [(½)^5(½)^{6-5}] \times [(6!)/(6-5)! (5!)]$$
$$= (½)^6 (6)$$

and

$$\text{for } 4/6: p = [(½)^4(½)^{6-4}] \times [(6!)/(6-4)! (4!)]$$
$$= (½)^6 (15)$$

The sum for the entire tail

$$(6/6 + 5/6 + 4/6)$$
$$= (½)^6 + (½)^6 (6) + (½)^6 (15)$$
$$= (½)^6 (1 + 6 + 15)$$
$$= 15.62 (10^{-3})(22)$$
$$= .34$$

Doubling this value for the tail comes to 0.1156; this is greater than 0.05; therefore we cannot reject the null hypothesis.

These calculations are manageable for small panels, but as the number of observations becomes large, the binomial distribution begins to resemble the normal distribution (Lawless and Heymann, 1997). Before calculators and computers, these calculations became so cumbersome that Roessler et al. (1978) published tables that use the binomial expansion to calculate the number of correct judgments and their probability of occurrence. The use of the tables (see Table E.1 in Appendix) simplifies the procedure of determining if a statistical preference for one sample over another is significant in paired preference tests.

For example, in a paired preference test for the two lemonade formulations described above with formulation A prepared using the lemonade base and lemonade formulation B from freshly squeezed lemons, 50 consumers participated. Thirty consumers preferred lemonade B and 20 preferred sample A. According to Table E.1, a formulation would have to be preferred by 33 or more of the 50 panelists to be significantly preferred at the 5% level, and by at least 35 of the 50 panelists to be significantly preferred at the 1% level of significance. Because one sample was preferred by only 30 of 50 panelists, the panelists did not have a significant preference for one lemonade formulation over the other. If 34 or 35 consumers had indicated a preference for formulation A, the conclusions are that A was preferred over B at the 5% or 1% levels of significance, respectively.

The Chi-Square (χ^2) Test

The chi-square test is used to test hypotheses about frequency of occurrence. The binomial test is used to test whether a panel of consumers prefer lemonade sample A to lemonade sample B. The chi-square may be used to obtain the same type of information, but it can be used to test hypotheses in more than two categories (O'Mahony 1986). The chi-square statistic can be calculated from the following formula:

$$\chi^2 = S = \frac{(O - E)^2}{E}$$

Where

    O = the observed frequency, and

    E = the expected frequency

for a paired preference test, the general formula can be written as

$$\chi^2 = \frac{(O_1 - E)^2 - 0.5}{E_1} + \frac{(O_2 - E_2)^2 - 0.5}{E_2}$$

E_1 is equal to the total number of observations, n, times probability, p, of a choice for sample 1 by chance alone in a single judgment where $p = .5$ for paired preference test.

E_2 is equal to the expected number of choices for sample 2 or the total number of observations, n, times probability, p, of choices for sample 2 by chance alone in a single judgment where $q = 1 - p$, or $= 1 - .5$; therefore $q = .5$ for paired preference tests.

The continuity correction factor $-.5$ is needed because the χ^2 distribution is continuous and observed frequencies from preference tests are integers. It is not possible for one-half of a person to have a preference, and so the statistical approximation can be off by as much as $\frac{1}{2}$, maximally (Lawless and Heymann, 1997).

As in the binomial test, the alternative hypothesis determines whether a one- or two-tailed test is used. We use a two-tailed test because we have no knowledge of which sample, A or B, will be preferred should the H_0 be false. We also look up the appropriate value for the degrees of freedom (df) for the test.

$$
\begin{aligned}
(df) &= \text{Number of categories} - 1. \\
&= (\text{Sample A vs. Sample B}) - 1 \\
&= 2 - 1 \\
&= 1
\end{aligned}
$$

The paired preference test determined whether one product is statistically preferred over a second product; therefore, the df is equal to one. An χ^2 table (Table E.5 in Appendix) using $df = 1$ should be used.

In the previous example, where 60 or 100 consumers preferred the freshly squeezed lemonade, A, over the experimental lemonade prepared from a base mixture,

$$
\begin{aligned}
\chi^2 &= \frac{(60 - 50)^2 - 0.5}{50} + \frac{(40 - 50)^2 - 0.5}{50} \\
&= 1.99 + 1.99 = 3.98
\end{aligned}
$$

Our value is 3.98. We select the appropriate column under Table E.5. For $df = 1$ and probability of 0.5 or 0.1 that χ^2 is greater than or equal to the critical value of chi-square. These values are 3.84 and 6.64 at the 0.05% and 0.01% levels of significance, respectively. Our value is greater than 3.84 (the largest value to be significant at the 5% level, therefore we reject H_0. We reject the idea that equal numbers of consumers would prefer products A and B. There

---

PAIRED PREFERENCE TEST

Name _____   Session Code _____   Date _____

Please rinse your mouth with water before starting.

In front of you are two samples. Beginning with the sample on the left, test each one and **circle the sample that you prefer or circle no preference if that is your feeling about the samples.**

You may retaste as often as you need to.

      529   **648**   **No Preference**

Thank you

Figure 2.2   An example of a ballot to be used in a paired-preference test with the no-preference option.

were more who preferred A. If the value we obtained were equal to 3.84, we would still reject H_0 (O'Mahony 1986). We did not reject H_0 at the 1% level, where the critical value is 6.64.

Power of the Binomial and Chi-Square Tests

Using the same example, we rejected H_0 with the chi-square test but did not reject it using the binomial test. The chi-square test is more likely to reject H_0 than the binomial test; it is a more powerful test. What test is better to use? If conflicting results are obtained when using the chi-square and binomial methods of analyses, it is advisable to conduct more tests until the answers become clearer. O'Mahony (1986) discussed this issue as follows: "If you are rejecting a null hypothesis, it is better to be able to do so on both tests. If you wish to be completely confident about rejecting H_0, it is safer to reject it on the test that is likely to do so—the binomial test. If you are accepting H_0, it is safer to accept it on the test that least likely to accept it—chi-square."

Nonforced Preference

The data analyses described above are based on a forced choice paired-preference test. If the project leader and requester jointly decide on use of the "no preference" option (see Figure 2.2 for a sample scoresheet), this could provide useful information regarding the product. If this option is used, several choices for handling the data exist. The first option is to ignore the "no-preference" responses. Ignoring these results in a decrease in the number of panelists and a decrease in the power of the test. Another option is to split the "no-preference" responses equally. This assumes, mistakenly, that the panelists would have selected one of the product samples over the other if forced to make a choice. The sample size is not decreased, and there appears to be no effect on the power of the test. This procedure actually dilutes the signal-to-noise ratio by assuming that panelists who express no choice would respond randomly (Lawless and Heymann, 1997). This conservative approach protects from a false-positive result while still running the risk of missing a significant preference. Another alternative is to split the "no-preference" responses proportionately corresponding to ratio of preference between the two product samples (Odesky, 1967). This is not a common practice, and the risks associated with this procedure are not known. Although, when faced with this situation, it suggests that the test should have been a "forced choice" test. When the panel size is over 100 and fewer than 20% respond "no-preference," the confidence intervals for the multinomial distribution can be calculated. If the confidence intervals of the proportions of those expressing a preference do not overlap, one can test whether one product is preferred over the other.

$$\text{Confidence limits} = (X^2 + 2x) \pm \frac{(X^2 \, [X^2 + 4x(N-x)/N])^{1/2}}{2(N = X^2)}$$

where

   X = number of observed preference votes in a sample
   N = sample size
   $X^2 = 5.99$ at an alpha = 5% and $df = 2$

Misuse of the Paired-Preference Test

The paired-preference test should not be combined with discrimination tasks. The asking of preference questions should not follow a difference question. The consumers who are recruited to participate in the preference tests must be naive users of the product and will not qualify as panelists in discrimination tests. Members of a descriptive panel do not represent the target population of a food product sample. A discrimination test is an analytical test, whereas the preference tests measure hedonic ratings.

Rating Tests

Scale ratings reflect consumer panelists' perceived intensity of a specified attribute under a given set of test conditions. Various rating scales, the hedonic

Panelist Code _____

Session No. _____

Take a piece of sample 452 using a fork and place the whole piece into your mouth.

Please answer the following questions by completely filling in the square that
best reflects your feelings about this sample.

OVERALL, how would you rate this sample?

Dislike Extremely	Dislike Very Much	Dislike Moderately	Dislike Slightly	Neither Like nor Dislike	Like Slightly	Like Moderately	Like Very Much	Like Extremely
□	□	□	□	□	□	□	□	□

Figure 2.3   An example of a hedonic scale for overall acceptance.

scale, and the food action scale have been developed and used in hedonic tests
and food action rating.

Hedonic Test

The 9-point hedonic scale (Figure 2.3) is a rating scale that has been used for
many years in sensory evaluation in the food industry to determine the acceptance of a food and to provide a benchmark on which to compare results. Its
use has been validated in the scientific literature (Stone and Sidel, 1993). The
number of scale categories that have been used include the 5-, 7-, and 9-point
scale. The 3-point scale is not recommended for use with adult consumers,
because adults tend to avoid using the extreme points of the scale in rating
food product samples. However, in some instances, adaptations of the 9-point
hedonic scale in the form of a 9-point facial scale was found useful in testing
children's responses. The 3-, 5-, and 7-point facial hedonic scales (Fig 2.4)
were found to be appropriate for 3-, 4-, and 5-year-old children, respectively
(Chen et al., 1996).

Food Action Rating Scale (FACT)

The FACT rating scale (Table 2.2) was devised by Schutz (1965) to measure
acceptance of a product by a population. It is a measure of general attitude
toward a product. The eating scale includes action- as well as affective-type
statements. Nine categories are represented. One or more samples may be
tested. Samples are presented sequentially in a balanced order, and the panelist
is asked to decide which of the statements on the scale best represents his or
her attitude. The ratings are converted to numerical scores to facilitate statistical
analysis of data.

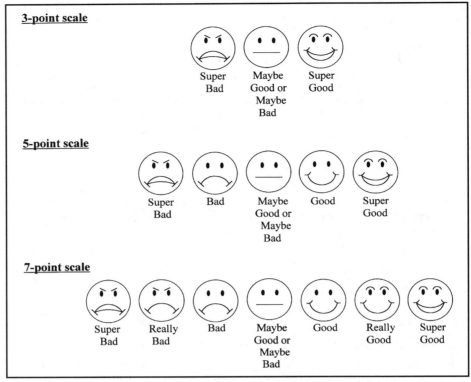

Figure 2.4 Examples of pictorial scale, 3-, 5-, and 7-point hedonic scales suitable for young children (Chen et al., 1996).

Table 2.2 Descriptors Used in the Food Action Rating Scale (FACT) Rating Devised by Schutz (1965)

I would eat this food every opportunity I had.

I would eat this very often.

I would frequently eat this.

I like this and would eat it now and then.

I would eat this if available but would not go out of my way.

I don't like it but would eat it on an occasion.

I would hardly ever eat this.

I would eat this only if there were no other food choices.

I would eat this only if I were forced to.

Preference Test—Ranking

Name _____   Session Code _____   Date _____

Please rinse your mouth with water before starting, before each sample and anytime you need to.

In front of you are five samples. Beginning with the sample on the left, taste each one.

After you taste all samples, you may retaste as often as you need to.

Rank the samples from the most preferred (= 1) to least preferred (= 5)

Sample	Rank (1 to 5) Ties are NOT allowed
643	_____
296	_____
781	_____
528	_____
937	_____

Thank you

Figure 2.5   An example of a ballot to be used in a ranking test.

Other Methods

Ranking Test

Ranking is, in effect, an extension of the paired-preference test. Many of the advantages of the paired-preference test apply to ranking. These include simplicity of instructions to participants, a minimum amount of effort to conduct, uncomplicated data handling, and minimal assumptions about level of measurement, as the data are treated as ordinal (Lawless and Heymann, 1997). Three or more coded samples are presented simultaneously, sufficient in amount so that the panelist can retaste the product. The number of samples tested is dependent upon the panelists' span of attention and memory as well as physiological considerations. With untrained or naive panelists, no more than four to six samples may be included in a test (ASTM, 1996). The panelist is asked to assign an order to the samples according to his or her preference. As with the paired-preference method, rank order evaluates samples only in relation to one another. An example of a scorecard for a ranking test is presented in Figure 2.5.

Ranking forces the consumer to make a judgment between samples. Unfortunately, rank-order can only give relative preference among samples tested at any one time (Peryam and Pilgrim, 1957). The amount of liking (or disliking) for individual samples cannot be adequately determined by this method (IFT/SED, 1981). A highly ranked sample does not necessarily correspond to a high acceptance rating. The simplicity of ranking makes it an appropriate consideration in situations where participants would have difficulty understanding scaling instructions, such as working with groups of people who cannot read, young children, across cultural boundaries, or in linguistically challenging situations (Coetzee and Taylor, 1996).

Ranking tests have several limitations. These include the fact that all products must be tested before a judgment can be made. This may result in sensory fatigue when a large number of products are evaluated and interactions occur from carryover effects of flavors. Consumer panelists, being untrained, may not understand or perceive the attribute being ranked. Finally, no indication of the absolute intensity (high or low) of the attribute being evaluated is given, and there is no measure of the magnitude of difference between products (Stone and Sidel, 1993). For these reasons, rating scales are preferred. However, if product preference is the only information needed, ranking can be useful and easy to perform (Colwill, 1987).

Analysis

The data are ordinal in nature, and the rank values are not independently distributed. The data are nonparametric and may be analyzed either by using the Basker tables (1988) or Friedman's test (Gacula and Singh, 1984). When using the Basker tables, panelists are forced to make a choice and are not allowed to have ties in the ranks.

Ranked data are quickly analyzed using the tables by Basker (1988). In the example below (Table 2.3), numerical values were assigned to each product by each panelist. The most preferred product was ranked 1 and the least preferred was ranked 5. Each consumer did a ranking of the samples once. A total of 103 consumers ranked the samples. In this example, rank totals for the seven products were calculated.

The Basker tables (Table E.2 in Appendix) indicate the critical difference for 103 consumers and five products is 62. The results are as follows:

For product A = 262 a
B = 291 ab
C = 315 abc
D = 351 bc
E = 356 c

Rank totals not followed by the same letter are significantly different according to this test. Product A is not significantly preferred over product B or C but

Table 2.3 Frequencies and Rankings for Poultry in Different Packages (1 = Best).

Package type and identification	Rank (W)	Frequency (v)	Weighted score (rank × frequency)	Rank totals
Resealable pouch (A)	1	40	40	232
	2	25	50	
	3	16	48	
	4	16	64	
	5	6	30	
Clear tray (B)	1	12	12	291
	2	28	56	
	3	36	108	
	4	20	80	
	5	7	35	
Clear box (C)	1	15	15	315
	2	18	36	
	3	31	93	
	4	24	96	
	5	15	75	
Dual-ovenable box (D)	1	19	19	351
	2	16	32	
	3	9	27	
	4	22	88	
	5	37	185	
Styrofoam tray (E)	1	17	17	356
	2	16	32	
	3	11	33	
	4	21	84	
	5	38	190	

is significantly preferred over products D and E. Product B is significantly preferred over product E.

The calculations based on the Friedman's test are discussed in Lawless and Heymann (1997). Using the Friedman's test, the formula to test whether there are differences is based on the chi-square statistic is

$$\chi^2 = \{12/[K(J)(J + 1)]\}\ E(Tj^2) - 3\ (K)(J + 1)$$

where

K = number of panelists

J = number of products, and

Tj = rank sums

With degrees of freedom for $\chi^2 = (K - 1)$.

When it has been determined that $\chi^2 = (K - 1)$.

When it has been determined that χ^2 is significant, a comparison of rank is done, Lawless and Heymann (1997) refer to this as the "least significant ranked difference" (LSRD), given by the formula:

$$LSRD = t[(NK(K + 1)/6]^{\frac{1}{2}}$$

where

    K = the number of samples

    N = the number of panelists

    t = the critical t-value at $df = N - 1$ and $\alpha = 0.05$.

From the χ^2 table, the critical value for $\alpha = 0.05$ and df, $N - 1 = 4$ is 9.49.

From the Friedman's test, the χ^2 value is greater than the table value of 9.49. Therefore, the preference rank differs significantly at the 5% level. To determine which products were ranked significantly higher than another, the LSRD is calculated as indicated earlier. For this example, LSRD = 2.78 $(103*5*6/6)^{\frac{1}{2}} = 63.08$. The results are as follows:

    For product A = 262 a

                B = 291 ab

                C = 315 abc

                D = 351 bc

                E = 356 c

Products with rank totals that are not followed by the same letter are significantly different. In this case the same results were obtained. However, the critical value arrived at using LSRD is slightly larger than those from Basker's tabulated values. The results from the Basker's tables are more conservative than those from the Friedman's test.

Ratio Scaling/Magnitude Estimation

Ratio scaling was proposed by Moskowitz (1974) as a method to quantify acceptance and preference. It is a technique for scaling that allows the respondent to use any positive number, so that the ratios between the numbers reflect the ratios of magnitude of the sensation that has been experienced. For example, if peanut sample 467 is given an overall acceptance rating of 50 and peanut sample 852 is liked only half as much, sample 852 is given a magnitude of

25 for overall acceptance. There is no physical scale that appears on the ballot, but the method involves a short orientation period for the consumers on the scaling process.

Two types of procedures are used. One method requires the presentation of a reference or anchor, which is generally assigned a fixed value for the numerical response. All subsequent stimuli are rated relative to this reference, which is called a "modulus." The other method does not require a modulus, and the consumer is free to choose any number for the first sample. The following ratings are based on the first number. Magnitude estimation and guidelines for data analysis are found in ASTM (1995) Standard Test method E 1697-95.

Magnitude estimation has been applied to a wide range of products but is not as frequently used as the paired-preference tests and tests employing the 9-point hedonic scale. ASTM (1996) indicated its usefulness in evaluation of moderate to large suprathreshold differences stating that the measurement of very small differences among similar products is more efficiently accomplished using other sensory techniques. The method of magnitude estimation may, for practical purposes, be used with consumers and even children (Collins and Gescheider, 1989). The data are a bit more variable than other bounded scaling methods, especially in the hands of untrained consumers (Lawless and Malone, 1986). Efforts to demonstrate that magnitude estimation is a more useful scale for measuring product acceptance or preference have also proven to be less than successful (Stone and Sidel, 1993). When scaling like or dislike using magnitude estimation, one can either use a unipolar scale for liking and have the panelist indicate whether the number represents liking or disliking, or use a bipolar scale with positive and negative numbers and a neutral point (Pearce et al., 1986). In unipolar magnitude estimation, only positive numbers are allowed, and the lower and upper ends of the scale represent not liking and liking, respectively (Moskowitz and Sidel, 1971; Giovanni and Pangborn, 1983). With the unipolar scale, the question should be asked whether the unipolar scale is a sensible response for a consumer panelist, as it does not recognize liking, disliking, or a neutral response (Lawless and Heymann, 1997). Lawless and Heymann (1997) state that the unipolar scale would make sense in consumer testing if items in the test were to be rated as all liked or all disliked. A bipolar scale, such as the 9-point hedonic scale, may be more appropriate.

Data-Collection Methods

Questioning and Observation Methods

Consumer sensory evaluation involves the collection of primary data from consumer panelists for decision making (ASTM, 1979). The type of data collected is defined by the objectives of the test. Data collection methods fall into two types of categories: (1) questioning, and (2) observation.

Questioning can be either by interview or questionnaire. The interview can be by telephone, personal contact, or by mail. The questionnaire can either be self-administered or questions may be asked by a trained interviewer, who also records answers to consumer responses.

The other method of data collection is by observation. Observation may be done directly by personal observers or indirectly through electronic or mechanical devices.

Data Collection by Questioning

Telephone Interviews

Telephone interviews are often used when collecting responses to questions in home-use tests. The advantage of collecting data through telephone interviews is that they are faster to conduct and may be less expensive to conduct than the personal interview because they do not require travel to consumers' homes. Recruitment of a statistically reliable sample of consumers may be facilitated when telephone interviews rather than personal interviews are conducted. Use of the telephone will allow the interviewer to easily recontact individuals who do not answer when the first telephone call is made. Sample questionnaires may be mailed to the consumer panelists ahead of time to facilitate telephone interviews.

The disadvantage of the telephone interview is its limited use because only verbal responses may be obtained. For example, it is not possible to conduct a sensory evaluation test during a telephone interview, and, instead, panelists usually recall their responses to the product. Furthermore, telephone interviews do not allow observation of the environment or the respondent. Another drawback to planning the use of telephone interviewing methods for data collection is the increasing number of unlisted telephones. The necessity for obtaining a representative sample becomes problematic when the target consumer population involves a large proportion of respondents with unlisted telephones. The problem of unlisted telephone numbers can be minimized by using random-digit dialing techniques when recruiting panelists for a home-use test.

Personal Interviews

The personal interview can be conducted in a sensory laboratory, central location, or in the respondent's own home during a home-use test. This method allows the collection of responses from consumers who may not have a telephone, who object to being interviewed by telephone, or have an unlisted number and cannot be reached when using the telephone directory for recruitment. Personal interviews, unlike telephone interviews, allow sensory evaluation tests to be conducted during the interview, and allow the interviewer to use visual aids and different types of scales to aid in the interview process. It allows the interviewer to observe and evaluate the testing environment, such as the home or office of the respondent. In addition, it is possible for an observer to view a personal interview as it is being conducted. Therefore, more information can be obtained in a personal interview than in a telephone interview. Although the respondent may walk out during the interview, it is more difficult for the respondent to terminate a personal interview than a telephone interview.

The major disadvantage of the personal interview as a data-collection method is the difficulty of obtaining a statistically reliable sample (ASTM, 1979) to interview. Personal interviews often involve travel for either the interviewers or the respondents. If the interviews take place in consumers' neighborhoods, security or safety may be a problem for the interviewer. Personal interviews are more time consuming than the telephone interview and therefore may incur a variable and higher cost per interview.

Mailed Questionnaires

The elimination of interviewer bias and the lower cost of data collection are the main advantages of this data-collection method over the telephone and personal interviews. The method allows for a statistically, demographically balanced sample of consumers to be obtained easily. Consumer respondents may be more willing to provide confidential personal information in response to questions on a self-administered questionnaire than during a telephone or personal interview.

The disadvantages of the mailed questionnaire is that all respondents must be able to read to respond to the questions. In addition, return of the questionnaire is variable; it may take one to six weeks or not at all; therefore, response rates are lowest in mailed interviews compared to telephone and personal interviews. Furthermore, responses may be less truthful and require verification.

Data Collection by Observation

Observation methods are used less extensively than the questionnaire data-collection methods. The consumers may or may not be aware of what is being observed; thus, there will be less of a tendency on the part of consumers to change their behavior. The observation may be direct or indirect, through a piece of equipment or instrument.

The advantage of personal observation is that actual behaviors are observed; therefore, recall by the respondent is not necessary. Errors of recall and distortion of events are eliminated. Because there is no direct interaction between the observer and the consumer, bias due to interviewer effects and subject awareness ("halo-effect") are eliminated. If the test is video- or audio-taped, a documentary record of the observation is available until the tapes are destroyed. The method allows for some individuals to be singled out for in-depth questioning.

The disadvantages are that only specific behaviors are available for observation. Sampling biases may occur due to time and location. The observer may vary over time due to fatigue, confusion, or a shift in attention.

Other Types of Observation Methods

Measurements of consumption, use, wear or erosion, and accumulation have been used to determine acceptance. Examples of such measurements of consumption are the weight of potato chips consumed while watching a movie; counting discarded containers, wrapping, and packaging materials such as bottle caps from a machine dispensing multiple flavors; or noting counts of different products purchased in a simulated supermarket-setting test (Hashim et al., 1995). The advantages of these methods are that they are inconspicuous, and respondents are not aware of the measurement of behavior.

Factors Affecting Data Collection

Length of the Questionnaire

The length of the questionnaire is an important consideration and should be carefully designed by the project coordinator. The questionnaire should ask the minimum number of questions to accomplish project objectives. Acceptance tests usually last less than one-half hour. The questionnaire needs to be pilot-tested using a small sample of consumers prior to use to determine the appropriate number of questions and length of the test.

Scheduling of Interviews or Tests

If the interview or test will take longer than ten to twenty minutes, it is advisable to prerecruit consumer panelists and schedule interview appointments rather than to intercept them from the general public. If the test design involves a sample of consumers who work in their homes and have flexible schedules, interviews or tests may be scheduled during the regular working hours. Otherwise, interviewing or tests should be scheduled after work hours or in the evening to enable employed individuals to participate. In other cases, interviewing or testing may be conducted in the location where consumers normally use the product (ASTM, 1979).

Vacation schedules, school openings and closings, holidays, and special events should be considered when scheduling a test or interview. Exceptions to these are interviews or tests on products used during these special occasions or holidays. Interviews with homemakers should also be avoided when children or other distractions may affect the consumers' responses. Hours that are usually spent in meal preparation should be avoided when interviewing.

The Questionnaire

Questionnaire Design

Preparation of the questionnaire is initiated once the decision has been reached on the test method, experimental design, and other information that was considered important to the test objective. In most cases, the project leader develops the questionnaire.

Questionnaire design is started by asking and answering the following: the questions to be asked, the sequence of questions to be asked, and how each question should be worded. Questionnaire wording has been shown to influence the data collection; therefore, wording of the questionnaire is extremely important. Ambiguity should be prevented at all times. Consumers participating in these tests should not only understand the questions but also be able to answer the questions in the frame of reference intended (ASTM, 1979).

Rules on Question Structure and Wording

Some rules on question wording were proposed by various authors such as ASTM (1979) and Meilgaard et al. (1991), and have been adapted for use in this chapter. They are listed in Table 2.4.

Question Sequence

Questions should be carefully and strategically positioned. The placement of a question on the questionnaire may affect the usefulness of the information obtained. Furthermore, bias may result from the positioning of questions in a certain order. The general rule is to move from general questions to the specific. The most important questions should be asked first. This is to ensure that data are collected from the consumers before extraneous factors or boredom set in.

Other Considerations

Last, the type of scale used and the choices or number of categories on a scale need to be considered. Consumer respondents may tend to avoid either end of

Table 2.4 Rules on Question Structure and Wording

1. Keep the questions clear and similar in style. To avoid confusion, the direction of the scales should be uniform.
2. Direct questions to address differences that are detectable and can differentiate products.
3. Consider the importance of including a personal question such as "What is your family income?" Respondents may consider this question obtrusive and may not answer the question.
4. Overelaboration can produce contradictions.
5. Do not overestimate the respondent's ability to answer specific questions such as those involving recall or estimation. For example, "How much salt do you use when you prepare food?"
6. Avoid double negatives.
7. Questions that talk down to respondents should be reworded.
8. Questions should be simple, direct, and encourage consumers to respond.
9. Questions should be actionable.

a scale and mark the middle of the scale. Likewise, responses on multiple-choice questions should be balanced.

Demographic Questions

If demographic questions are asked in a questionnaire, they should not be asked initially but should be positioned at the end, after the most important research issues have been answered. When using consumers recruited from a database and their demographic information is available, asking demographic questions would not serve any purpose.

Open-Ended Questions

The use of open-ended questions is controversial and not recommended by Stone and Sidel (1993). However, others (Meilgaard et al., 1991) consider space on a scoresheet for open-ended questions desirable.

Pretesting a Questionnaire

The questionnaire should be pretested to eliminate ambiguous terms, confusing items, and cues. In addition, the pretest will help to ensure that the questionnaire is understood by the consumers, that they will have no difficulty following directions or completing the tasks within a reasonable length of time. Ideally, pretesting of the questionnaire on untrained individuals who possess the characteristics of the target population should be used. The pretest will determine whether the questionnaire is too long or poorly designed.

Final Considerations

Finally, in designing the questionnaire, it is important to consider coding and tabulation of the data for manual or computer data processing, and the analysis of the data and the statistical package to be used. Allowing for these will result in considerable savings of time and effort.

Types of Questions

The types of questions can be broken down to structure and scaling types. Structure of questions can be classified into unstructured (open-ended), semistructured, and structured.

Semistructured and Unstructured Questions

Both types are without a format. They allow the consumer respondents to answer in any manner they see fit. There are no suggested answers. The questions allow for greater scope and natural exploration. They are designed to identify problems or to search for new avenues of thought. They eliminate researcher bias; responses do not have boundaries. Respondents get to contribute more than the researcher has thought about. One can expect a wide variety of responses, as is the case in focus group discussions, or no responses, when used in a self-administered sensory evaluation questionnaire. In contrast, the most obvious responses or issues may not be mentioned.

The wide variety of answers makes coding, quantification, and interpretation difficult and time consuming. This contributes to the cost of the research. If a personal interviewer is asking the questions, the open-ended answers are subject to differences in interpretation and coding by different interviewers, and the interviewer is asked to write a verbatim transcription of the responses in addition to having an audiotaped record of the interview. In such cases, the researcher is required to summarize comments from the respondents. A large sample is required in order to make comparisons of data. In some cases, several replications of the testing are needed to determine reliability of the responses (Galvez and Resurreccion, 1992). Another disadvantage of having open-ended questions is that most people, and even more so, inarticulate people, do not respond. The recommendation to include unstructured or open-ended questions is controversial. Unstructured questions can provide useful information to the researcher and are recommended by ASTM (1979) as a necessary part of many questionnaires, including sensory evaluation questionnaires as a comment after the last question is asked. However, Stone and Sidel (1993) discourage its use on the basis of limited response and therefore its usefulness.

At any rate, when asking unstructured or semistructured questions, both alternatives of a question should be stated (ASTM, 1979). For example, "What do you like (or dislike) about the sample?"

Probing

Probing is continuing to question in an attempt to get more information or to clarify certain points already raised. Typical questions might be "Is there anything else that irradiated chicken makes you think about?" or "Could you tell us more about what you don't like about dark chicken meat?" Probing should be planned. For example, it should appear in strategic positions in a focus group moderator's guide. In addition, the moderator should be trained in the approach to be used and the intensity of probing. The moderator must avoid questions that suggest ideas not previously mentioned or that could lead the participants to feel or think a certain way.

Structured Questions

Structured questions are the most common type used to measure demographic characteristics, such as age, sex, consumer attitudes, and so on. The use of structured questions relies heavily on scaling theories. The four categories of scales used in questionnaire design are normal, ordinal or rank, integer or interval, and ratio scaling. Some types of structured questions are rank order, paired comparison, interval scales, magnitude estimation, or ratio scales.

Advantages

The advantages of using structured questions is that they allow for greater control of the interview and responses. The respondents themselves fill out the questionnaire's sensory or ratio scales. Interviewer bias is minimized because the consumers fill out their own responses on the questionnaire. Responses are easily quantifiable and allow for electronic data processing. The ability for electronic compiling and data processing will likely lower the cost of the research. It assumes that the respondents have prior knowledge about the product, such as through purchase and/or use, and researcher is aware of all possible responses to the question.

Disadvantages

The structured responses minimizes the breadth of possible answers that may be obtained from the test. The scale used may not correlate well with psychological

Table 2.5 Types of Scales

1. Nominal scales are used to denote membership in a category, group, or class.
2. Ordinal scales are used in ordering or ranking.
3. Interval scales are used to denote equal distances between points and are used in measuring magnitudes, with a zero point that is usually arbitrary.
4. Ratio scales are used in measuring magnitudes, assuming equality of ratios between points and the zero point is a "real" zero.

behavior. With the exception of magnitude estimation or ratio scales, the values cannot be easily compared (ASTM, 1979).

Types of Scales Used

Four different types of scales exist (Stone and Sidel, 1993). These are the nominal, ordinal, interval and ratio scales and are defined in Table 2.5.

Nominal Scale

A nominal scale can be related to the calendar on the wall or the baseball program where the players are numbered (ASTM, 1979). Numbers are used to label, code, or otherwise classify (Stone and Sidel, 1993) or differentiate items or responses. Nominal scales are often used in asking demographic questions such as gender (male, female) or employment classification (full-time, part-time, retired, not employed) from panelists without taking a substantial amount of time. The responses may be analyzed using frequency counts, distributions, modes (the response given by the largest number of panelists), and chi-square (Stone and Sidel, 1993).

Ordinal (Ranking) Scale

Ordinal scales either use numbers or words organized from "highest" to "lowest" and "most" to "least." These tell whether the item has a quality of being equal to, greater than, or less than. For example, "Which honey sample has more orange flavor?" Ranking is one of the most commonly used types of ordinal scale. Analysis of ranked data is straightforward. Simple rank-sum statistics can be found in published tables (Basker, 1988) that corrected the previous errors in the analysis of Kramer's rank-sum test. In addition, different methods include those appropriate for nominal scales, and, in particular, those methods classified as nonparametric methods can be used. These methods include the

Wilcoxon signed ranks test, Mann–Whitney, Kruskal–Wallis, Friedman's two-way analysis of variance, chi-square, and Kendall's coefficient of concordance.

Interval Scale

Rating scales are the most widely used scales in sensory evaluation. Their popularity may be attributed largely to their ease of use. In addition, a large number of statistical tests can be used to analyze results from tests utilizing rating scales.

This scale is one in which the interval or distance between points on the scale is assumed to be equal. It helps define the differences or magnitude between named or numbered items. An example of an interval scale is the 9-point hedonic scale. Interval scales are considered to be truly quantitative scales, and most statistical procedures, including means, standard deviations, t-test, analysis of variance, correlation and regression, and factor analysis are among the many statistical methods that may be used to analyze the data.

Ratio Scale

The ratio scale implies a constant ratio between fixed points or a zero point. Magnitude estimation is the most frequently used ratio scale used in consumer testing.

Criteria for Selection of Measurement Scales

Selection of a scale for a specific test is one of the most important steps in implementing a consumer test. Criteria that should be used in selecting or developing a scale for a sensory test are listed in Table 2.6.

Hedonic Scale

Among all scales and test methods, the 9-point hedonic scale has gained special consideration because of its suitability in measurement of product acceptance and preference. It was developed by Jones et al. (1955) and Peryam and Pilgrim (1957). Further research resulted in a 9-point bipolar scale, with a neutral point in the middle and nine statements that describe each of the points or categories (Figure 2.6). The scale is easily understood by panelists and is easy to use. Unlabeled check-box scales can be used in place of integer numbers to avoid bias. In these scales, there are no numbers or labels associated with the intermediate categories. These scales should carry verbal end labels to anchor the scale to common frame of reference (Lawless and Heymann, 1977). The reliability and validity of the 9-point hedonic scale in the assessment of several hundred food items has been confirmed (Peryam et al., 1960; Meiselman et al., 1974).

Table 2.6   Criteria for Selecting or Developing Scales

1. The scale should be valid. It should measure the attribute, property, or performance characteristic that needs to be measured as defined by the objectives of the study.
2. The scale should be unambiguous and easily understood by panelists. Questions as well as the responses should be easily understood by the panelists.
3. The scale should be easy to use. Consumer tests involve panelists who are not trained.
4. It should be unbiased (Stone and Sidel, 1993). Results should not be an artifact of the scale. Bias may result from the words or numbers used in a scale. Unbalanced scales result in biased results.
5. It should be sensitive to differences. The number of categories used and the scale length will influence the sensitivity of the scale in measuring differences.
6. The scale should consider endpoint effects.
7. The scale should allow for statistical analyses of responses.

Modifications to the 9-point scale have been suggested, such as eliminating the neutral point ("neither like nor dislike"), or simplifying the scale by eliminating options such as the "like moderately" and "dislike moderately" points on the scale, or truncating the endpoints by eliminating the "like extremely" and "dislike extremely" points, but they were either unsuccessful or had no practical value. These steps would bring in the problem of end-use avoidance or the hesitation of panelists to use the end categories. Truncating a 9-point scale to a 7-point scale may leave the consumer panelist with only a 5-point scale. It is therefore best to avoid the tendency to truncate scales in experimental planning (Lawless and Heymann, 1997). Another concern was the bipolar scale and the analysis of the results as unidirectional. However, there is no evidence that consumers experience difficulty with the scale and that the statistical analysis presents a problem (Stone and Sidel, 1993).

Stone and Sidel (1993) state that the stability of responses and the extent to which the data can be used as a sensory benchmark for any particular product category are of particular value. Acceptance of a product can be quantified by a mean and a standard deviation. When tested against a wide array of competitive products, an ordering of means of the different products can be made. It likewise allows the comparison of a product with the highest possible scores that the product category may achieve. Furthermore, parametric statistical analysis, such as analysis of variance (ANOVA) of acceptance data from tests using the 9-point hedonic scale can provide useful information about product differences. Stone and Sidel (1993), contrary to O'Mahony (1982) and Vie et al. (1991), note that data from this scale should not be assumed to violate the normality

For office use only
Panelist #

Panelist #_____

For office use only
Sample #

Sample #_____

Please consume all of your sample in order to evaluate all attributes.

Please remember one page of scoresheet is for one sample.

Please answer the following questions by completely filling in the square that best reflects your feelings about this sample.

1. OVERALL, how would you rate this sample?

Dislike Extremely	Dislike Very Much	Dislike Moderately	Dislike Slightly	Neither Like nor Dislike	Like Slightly	Like Moderately	Like Very Much	Like Extremely
□	□	□	□	□	□	□	□	□

2. How would you rate the COLOR of this sample?

Dislike Extremely	Dislike Very Much	Dislike Moderately	Dislike Slightly	Neither Like nor Dislike	Like Slightly	Like Moderately	Like Very Much	Like Extremely
□	□	□	□	□	□	□	□	□

3. How would you rate the FLAVOR of this sample?

Dislike Extremely	Dislike Very Much	Dislike Moderately	Dislike Slightly	Neither Like nor Dislike	Like Slightly	Like Moderately	Like Very Much	Like Extremely
□	□	□	□	□	□	□	□	□

4. How would you rate the TEXTURE of this sample?

Dislike Extremely	Dislike Very Much	Dislike Moderately	Dislike Slightly	Neither Like nor Dislike	Like Slightly	Like Moderately	Like Very Much	Like Extremely
□	□	□	□	□	□	□	□	□

Please turn to next page.

Figure 2.6   Example of a hedonic scale for overall acceptance, and acceptance of color, flavor, and texture of a sample. Continued on next page.

assumption. A plot of responses of 222 consumers who used the 9-point hedonic scale in evaluating twelve products resulted in a sigmoid plot, which indicates that the scores are normally distributed.

Consumer responses from use of a hedonic scale can likewise be converted to ranks or paired-preference data. To convert to paired-preference data, it is necessary to count the number of subjects who scored one product higher and analyze the result using $p = \frac{1}{2}$, or binomial distribution. The 9-point hedonic scale has yielded results that are reliable and valid. Efforts to improve the scale

Please consume all of your sample in order to evaluate all attributes.

Please remember one page of scoresheet is for one sample.

Please answer the following questions by completely filling in the square that best reflects your feelings about this sample.

1. OVERALL, how would you rate this sample?

Dislike Extremely	Dislike Very Much	Dislike Moderately	Dislike Slightly	Neither Like nor Dislike	Like Slightly	Like Moderately	Like Very Much	Like Extremely
□	□	□	□	□	□	□	□	□

2. How would you rate the COLOR of this sample?

Dislike Extremely	Dislike Very Much	Dislike Moderately	Dislike Slightly	Neither Like nor Dislike	Like Slightly	Like Moderately	Like Very Much	Like Extremely
□	□	□	□	□	□	□	□	□

3. How would you rate the FLAVOR of this sample?

Dislike Extremely	Dislike Very Much	Dislike Moderately	Dislike Slightly	Neither Like nor Dislike	Like Slightly	Like Moderately	Like Very Much	Like Extremely
□	□	□	□	□	□	□	□	□

4. How would you rate the TEXTURE of this sample?

Dislike Extremely	Dislike Very Much	Dislike Moderately	Dislike Slightly	Neither Like nor Dislike	Like Slightly	Like Moderately	Like Very Much	Like Extremely
□	□	□	□	□	□	□	□	□

Thank you very much for your time.

Figure 2.6   (*Continued*)

have been unsuccessful, and it should continue to be used with confidence (Stone and Sidel, 1993).

Demographic Information

Demographic information is usually collected during prescreening. If recruiting from a data bank of panelists, information in the database may be updated then. If recruiting consumer panelists by intercepting them or in face-to-face recruitment situations, the information is usually collected during the screening process. If it is not possible to collect demographic information before the test date, demographic information may be collected after the test is completed to reduce the time of the actual test.

References

Amerine, M. A., Pangborn, R. M., and Roessler, E. B. 1965. *Principles of Sensory Evaluation of Food,* M. L. Anson, C. O. Chichester, E. M. Mark, and G. F. Stewart, eds. Academic Press, New York. p. 250.

ASTM, Committee E-18. 1979. *ASTM Manual on Consumer Sensory Evaluation,* ASTM Special Technical Publication 682, E. E. Schaefer, ed., American Society for Testing and Materials, Philadelphia, PA, pp. 28–30.

ASTM. 1995. Standard test method for unipolar magnitude estimation of sensory attributes. In *Annual Book of ASTM Standards: Vol. 15.07. End Use Products.* American Society for Testing and Materials, Conshohocken, PA, pp. 105–112.

ASTM, Committee E-18. 1996. *Sensory Testing Methods.* ASTM Manual Series: MNL 26, 2nd ed., E. Chambers, IV and M. B. Wolf, eds. American Society for Testing and Materials, West Conshohocken, PA. pp. 38–53.

Basker, D. 1988. Critical values of differences among rank sums for multiple comparisons. *Food Technol.* 42(2):79, 80–84.

Chen, A. W., Resurreccion, A. V. A., and Paguio, L. P. 1996. Age appropriate hedonic scales to measure food preferences of young children. *J. Sensory Stud.* 11:141–163.

Coetzee, H., and Taylor, J. R. N. 1996. The use and adaptation of the paired comparison method in the sensory evaluation of hamburger-type patties by illiterate/semi-illiterate consumers. Food Qual. and Pref. 7:81–85.

Collins, A. A., and Gescheider, G. A. 1989. The measurement of loudness in individual children and adults by absolute magnitude estimation and cross modality matching. J. Acoustic. Soc. Amer. 85:2012–2021.

Colwill, J. S. 1987. Sensory analysis by consumers: Part 2. *Food Manuf.* 62(2):53, 55.

Gacula, M. C., Jr., and Singh, J. 1984. Statistical Methods in Food and Consumer Research. Academic Press, Orlando, FL.

Galvez, F. C. F., and Resurreccion, A. V. A. 1992. Reliability of the focus group technique in determining the quality characteristics of mungbean (*Vigna radiata* (L.) Wilczec) noodles. *J. Sens. Stud.* 7:315–326.

Giovanni, M. E., and Pangborn, R. M. 1983. Measurement of taste intensity and degree of liking of beverages by graphic scaling and magnitude estimation. *J. Food Sci.* 48:1175–1182.

Hashim, I. B., Resurreccion, A. V. A., and McWatters, K. H. 1995. Consumer acceptance of irradiated poultry. *Poultry Sci.* 74:1287–1294.

IFT/SED. 1981. Sensory evaluation guideline for testing food and beverage products. *Food Technol.* 35(11):50–59.

Jellinek, G. 1964. Introduction to and critical review of modern methods of sensory analysis (odour, taste and flavour evaluation) with special emphasis on descriptive sensory analysis (flavour profile method). *J. Nutrit. Dietet.* 1:219–260.

Jones, L. V., Peryam, D. R., and Thurstone, L. L. 1955. Development of a scale for measuring soldiers' food preference. *Food Res.* 20:512–520.

Lawless, H. T., and Heymann, H. 1997. *Sensory Evaluation of Food: Principles and Practices.* Chapman and Hall, New York.

Lawless, H. T., and Malone, G. J. 1986. A comparison of scaling methods: Sensitivity, replicates and relative, measurement. *J. Sens. Stud.* 1:155–174.

Meilgaard, M., Civille, G. V., and Carr, B. T. 1991. *Sensory Evaluation Techniques,* 2nd ed., CRC Press, Boca Raton, FL.

Meiselman, H. L., Waterman, D., and Symington, L. E. 1974. *Armed Forces Food Preferences,* Tech. Rep. 75-63-FSL. U.S. Army Natick Development Center, Natick, MA.

Moskowitz, H. 1974. Sensory evaluation by magnitude estimation. *Food Technol.* 28(11):16, 18, 20–21.

Moskowitz, H. R., and Sidel, J. L. 1971. Magnitude and hedonic scales of food acceptability. *J. Food Sci.* 36:677–680.

Odesky, S. H. 1967. Handling the neutral vote in paired comparison product testing. *J. Market. Res.* 4:149–167.

O'Mahony, M. 1986. Sensory Evaluation of Food. Marcel Dekker, New York.

Pearce, J. H., Korth, B., and Warren, C. B. 1986. Evaluation of three scaling methods for hedonics. *J. Sens. Stud.* 1:27–46.

Peryam, D. R., and Pilgrim, F. J. 1957. Hedonic scale method of measuring food preference. *Food Technol.* 11(9):9–14.

Peryam, D. R., Polemis, B. W., Kamen, J. M., Eindhoven, J., and Pilgrim, F. J. 1960. *Food Preferences of Men in the Armed Forces.* Quartermaster Food and Container Institute of the Armed Forces, Chicago.

Resurreccion, A. V. A. 1988. Applications of multivariate methods in food quality evaluation. *Food Technol.* 42(11):128, 130, 132–134, 136.

Roessler, E. B., Pangborn, R. M., Sidel, J. L., and Stone, H. 1978. Expanded statistical tables for estimating significance in paired-preference, paired-difference, duo-trio and triangle tests. *J. Food Sci.* 43:940–943.

Schutz, H. 1965. A food action scale for measuring food acceptance. *J. Food Sci.* 30:365–374.

Stone, H., and Sidel, J. L. 1993. *Sensory Evaluation Practices,* 2nd ed. Academic Press, San Diego, CA.

Vie, A., Gulli, D., and O'Mahony, M. 1991. Alternate hedonic measures. *J. Food Sci.* 56:1–5.

3

Test Procedures

Project Plan

The design, organization, implementation, and management of the consumer sensory test require meticulous planning in order to obtain the desired information. A properly designed consumer test increases the confidence with which a decision is reached. Planning for testing includes the definition of objectives, selection of an appropriate test and experimental design, identification and recruitment of the consumer sample, scheduling and implementation of the tests, screening of samples, selection of sample preparation method, data collection, data processing and analysis, interpretation, and reporting of results in a timely manner. Such planning takes time to develop but, when thoroughly done, quickly pays for itself in terms of consumer testing efficiencies, panelist participation, and credibility of results. Furthermore, organization of the tests should reflect realistic industrial practices (Stone and Sidel, 1985).

The very first step in conducting a consumer sensory test is to define the problem and identify critical objectives. Once the critical objectives have been clearly stated and jointly agreed upon by the sensory project leader and the technical or management individual or group requesting the research, the test can be designed.

All information regarding the consumer test plan should be retained in a file and include the items shown in Table 3.1. The written test request should be reviewed by both parties. If changes are necessary, the project leader needs to rewrite and submit these to the client for final approval. The test request will provide the foundation for the project plan. In most cases, a meeting between the client and project director is advisable and often necessary. During these meetings, the project leader and the client should discuss the different product samples thoroughly to ensure that the most appropriate affective test methods will be used. The characteristics of the consumer panel to be used during the test should also be discussed during the meeting. Although such a meeting may appear to some as superfluous and redundant, its usefulness in the design of the test more than compensates for the time and effort expended by both parties.

Table 3.1   The Project Plan

1. The name, address, telephone, fax, and E-mail numbers of the client.
2. A clear statement of the objectives.
3. A brief description of the test procedural sequence, including the interview with the client or requester of the test, a product review, if warranted, test schedule, and proposed reporting schedule and distribution of reports.
4. Examples of the test request and report forms.
5. A description of the consumer target population and panelist recruitment criteria, a copy of the recruitment script, recruitment screener, and description of how recruitment performance will be monitored.
6. Selected experimental designs, with notes on their use.
7. A brief description of each of the methods to be used. Additional information in the file should include notes on the questionnaire design and the ballot for each test, the method and a detailed description of each procedure, including preparation of samples, suggested containers, and serving procedures.
8. Identification of test samples, number of test products, and number of responses per sample.
9. Guidelines on test procedure, including sample coding procedures, instructions on serving, serving size, timing, lighting, and other considerations, such as quality assurance for the test.
10. Methods for data processing, analysis, and data interpretation.
11. Suggestions for panelist incentives.

Such a project plan will enable the project leader to identify which tasks need clarification, minimize duplication of effort, and help to delineate areas of responsibility and appropriate decision points. A checklist (Appendices A–D) for each consumer test should be prepared to ensure that material in the file that covers the factors to be used, and to make sure that tasks are completed in the time required for the test. Such a checklist will be helpful to the project leader to keep track of completion of tasks and increase the probability that the test will proceed smoothly. The following sections describe planning issues on design, organization, implementation, and management, and provide additional information regarding test procedures.

Requests for Assistance

Client requests for assistance may be oral or written. A written client request is an essential document for the project plan and guides all test activities. Oral requests should be transformed to written requests, dated, and placed in the file. The request form should consist of no more than one page. Both the client and the project leader will provide input for different portions of the form.

The request form should include the following information: the client's name, address, telephone, fax numbers, and e-mail address, test objective, product identification (name of products; product code; how, when, and where manufactured), amount of product available, location, preparation, specifications if known, individuals to be included and excluded from the panel, and report distribution. In addition, the project leader completes such items such as receipt date, type of test method requested, suggested test dates, design, number and type of panelists, sample presentation, number of replications, serving conditions, sample quantity, sample temperature, carrier, serving containers, lighting conditions, and other conditions. Comments of the experimenter may also be included in the form. A meeting may be needed to complete the form.

In other cases, the client request may be discussed by both the project leader and the client. The project leader writes a preproposal containing the following items: the name of the client, mailing address, telephone and fax numbers and e-mail address; test objective; product description and identification, amount of product available, location, specifications; the composition of consumers to be recruited for the test and characteristics of consumers to be excluded; the type of report, whether preliminary, one-page or final report, dates due and distribution. In addition the test method requested, suggested test dates, design and number of panelists, number of replications, serving conditions, serving container sample volume or weight, sample temperature, carrier if any, and lighting conditions. All the above should preferably be on one page. The preproposal is usually prepared from oral, telephoned, or face-to-face requests from the client; therefore, it needs to be faxed to clients with instructions to affix their signature and approval on the preproposal. The signed preproposal should be part of the project file. A preproposal format is in Figure 3.1.

Objectives

A clear statement of the objectives of the test is an essential first step in the planning and success of a project. The design of consumer affective tests depends, first and foremost, on the test objectives. The objectives provide the basis for formulation of hypotheses, experimental design, the specific affective test methods, and, finally, the statistical analysis methods that will be used. The statement of test objectives may consist of a general objective and one or more specific objectives, and should be formulated by both the client and the project leader. A written test request should be reviewed by both parties to determine whether the objectives are clearly stated and reasonable. It is of utmost importance that there be agreement between the project leader and the client on the objectives. If changes are necessary, the project leader needs to

---

PREPROPOSAL SUBMITTED TO:

Name of Company

Date

Name, Mailing Address, Telephone and Fax numbers, and E-mail address of Client.

Title: (STATE BRIEF TITLE OR PROJECT)

Project Leader: (Name, mailing address, telephone and fax numbers, E-mail address)

Objectives:

Sample: (Sample name, description, sample amounts and availability, sample location, description and storage history, dates of delivery, etc.)

Design: (Treatments, replications, sensory replications, etc.)

Test method:

Sample presentation: (Presentation scheme, serving containers, etc.)

Serving conditions: (Sample temperature, sample amount, carrier, if any)

Environmental Conditions: (Lighting, etc.)

Panel: (Number of subjects, criteria for recruitment (including panelists to be excluded from testing):

Report: (Type, due date, report distribution)

Approvals: (Signature and date of client)

Contact Person:

Name

Mailing address

Telephone:　　　　　　　**Fax:**　　　　　　　**E-mail:**

Figure 3.1 Preproposal format for a consumer sensory test.

rewrite and submit these to the client for final approval. Specific objectives are sometimes included to provide more information. The number and extent to which these are carried out will depend on their importance in answering the general objective. It is recommended that specific objectives be listed in order of their importance. Some of these may be eliminated prior to implementation of the project because of sample size and specifications, time, cost, and experimental design constraints or other factors. A hypothesis should be formulated from the objectives. The project should be designed to answer the stated objectives.

Test Selection

The next step is the selection of the test method. The selection of a method depends primarily on the objectives of the project. The test design, selection of a method for analysis, and interpretation of results are responsibilities of the project leader. Delegating this important responsibility solely to a statistician who is not congnizant of the subtleties of sensory evaluation, or other individual, such as the client, is not recommended (Stone and Sidel, 1993). For this reason, the project leader must have sufficient knowledge about statistical analysis procedures to recognize the need for assistance from statisticians and other sources of technical assistance when these occasions arise. If the project leader can foresee the need for statistical consultation, a meeting with the statistician should be conducted as early as possible during the early stages of project planning and continued throughout the course of the project. The different methods used in affective tests were described in Chapter 2.

Psychological Errors

In the selection of the test method, experimental design, panel recruitment, sample presentation scheme, and analysis and interpretation of consumer test results, it is important to know about the different sources of psychological error, which in turn can influence individual responses. Sidel and Stone (1993) listed and discussed a number of psychological errors, some of which can have a profound effect on the measurement of consumers' responses. Careful product selection, the use of qualified consumer test participants, and in certain cases, asking fewer questions will minimize the errors but not cancel them. All findings must be examined for these errors. The following list of errors includes those discussed by Stone and Sidel (1993) that apply to affective tests:

1. Error of central tendency
2. Sample presentation-order error
3. Expectation error
4. Stimulus error
5. Leniency error
6. Halo effect
7. Contrast and convergence errors

Error of Central Tendency. It has been suggested that consumer panelists may avoid the extreme points in a scale, such as the like extremely (= 9) or the dislike extremely (= 1) points on a 9-point hedonic scale (Jones et al.,

1955; Peryam and Pilgrim, 1957), and score in the midpoint of the scale to effectively reduce the scale to a 7-point scale. This error is more likely to occur when the panelists are either unfamiliar with the test methods or the test products and is less prevalent in tests wherein trained panelists are used. Although the recommendation was made by Amerine et al. (1965) that this error could be minimized by changing the words or their meaning, or by spacing the terms farther apart, Stone and Sidel (1993) state that this approach is more applicable to scales other than the 9-point hedonic scale, which does not result in this error when using a balanced serving order, and when the panelists are oriented as to the type of scale to be used prior to a test. This error should be considered when the requester of the test or client recommends reduction of a 9-point hedonic scale by two categories. This practice effectively reduces the scale to a 7-point scale that is less sensitive and is therefore not recommended.

Sample Presentation Order Error.   This error is also referred to as a time-order error (Stone and Sidel, 1993), order effect, or position effect and results in the first sample being scored higher than expected. The problems in interpretation of the results are minimized by using a balanced sample presentation scheme and interpreting results as a function of serving order. Different scenarios that illustrate this error and its interpretation are explained by Stone and Sidel (1993).

Expectation Error.   When panelists' knowledge about a product influences their responses due to their expectation of a specific attribute, this is said to result in expectation error. For example, employees of quick-serve restaurants who are asked to participate in a consumer test involving a new precooked, frozen, and reheated hamburger may expect the test hamburgers to have an off-flavor and therefore score these lower than they ordinarily would. This is only one of the many reasons why employees, as well as panelists who have a high degree of knowledge about a product, should not be recruited to participate in consumer affective tests regarding the products.

Stimulus Error.   When the participants in an affective test have any actual or perceived knowledge about products being tested, results obtained may not be as expected. For example, during the formulation and subsequent testing of a carbonated beverage, persons involved in the formulation may rate a beverage with more sweetener as more acceptable than the beverage that they perceive has less sweetener. To minimize the risk due to this error, the sample should be a representative sample and preferably be from one container as opposed to several containers of product. Stimulus error is a major reason why individuals who are directly involved in any of the product formulation steps for the product should not be involved in affective tests on the product.

Leniency Error.   When panelists rate the products on the basis of their feelings toward the group or individual that developed the product, or the person requesting the test rather than on the basis of their feeling toward the product, this is said to result in leniency error. This error can be minimized by use of consumer panelists who are unbiased in that they will not rate the product higher or lower because of positive or negative feeling about the product specialist in charge of the product, the project leader, or the interviewer. This type of error would be more likely to occur when using company employees than when using actual consumers.

Halo Effect.   This source of error is defined as the effect of one response on succeeding responses (Stone and Sidel, 1993) and has been observed in tests involving naive consumers, but not in those involving trained panelists. Lawless and Heymann (1997) defined the halo effect as the tendency to view a product more positively than normal due to one or more overriding or influential sensory attributes, or other positive influences, or the tendency of a sensory attribute to be rated as more intense or more hedonically positive due to other logically unrelated sensory attributes in a product.

The order in which the overall preference question is asked does not have an effect in minimizing the halo effect. The halo effect has been observed to occur when there are a large number of questions that need to be answered. To minimize error due to the halo effect is to either minimize the number of questions in a test or recognize that the halo effect is a source of error and interpret the results accordingly.

Contrast and Convergence Errors.   The problem of contrast and convergence is serious. An example is when a product with a lower priced ingredient is being substituted for the currently used ingredient and the two products are scored differently from each other, and the scores are more divergent than predicted; the product with the currently used ingredient is scored much higher. Convergence, on the other hand, is the suppression or masking of smaller differences between samples in the test, brought about by a large divergence between two or more products. Contrast and convergence errors are anticipated in such instances when an experimental product is tested against a commercial product, or when the identity of one or more products is known to the panelists. It is therefore important for the project leader to examine each product, long before the actual test, to determine the potential for contrast and convergence errors.

Experimental Design

In a consumer affective test, the experimental design is the written plan that designates the serving order of the products to panelists. The design of the experiment has a direct bearing on the conclusions that can be drawn from the test. Decisions should include consideration of the experimental design, sample treatments, test project variables, replications, number of samples to be tested, the type of comparisons (monadic, paired, and multiple-sample test designs) and the plan for presentation of samples that will be used. Power and sensitivity of the test should be major considerations, along with resources available for the test. A number of texts on experimental designs exist, including those by Box et al. (1978) and Cochran and Cox (1957).

The independent and dependent variables should be defined. The sample treatments should be identified, as should project variables. Levels of the test variables to be studied need to be determined. A statement of hypothesis will be made; both the null and alternate hypothesis will be determined. The level for alpha set and the level for beta risk and power will be considered.

The blocking structure can be determined next, with sessions and groups decided upon and treatment levels assigned to blocks. The number of judges will be determined along with sensory replications and observations needed. In analytical tests, replication is necessary.

In affective tests, responses are subjective, and large numbers of respondent are used. For this reason, replications of the test using another consumer panel are usually not conducted. The larger sample size provides better estimates of the population and validity to generalizations (Stone and Sidel, 1993). However, when testing new food product formulations or food products, it is important to test processing replications of the product sample. Testing of product replications involves testing of product samples from different batches. Examples of processing replication for a products are (1) for a "light" roasted peanut snack, use a different batch of peanuts and defat, roast, and coat each batch of peanuts separately for each replication; (2) for a snack chip formulation, prepare blends of composite flours separately and process these independently for each replication.

Stone and Sidel (1985) listed guidelines to consider in the design of experiments. These are summarized as follows:

1. Use balanced-block designs that minimize the interactions of subject, products, and time.

2. When testing four or fewer products, all products should be evaluated equally often by each subject.

3. All products should be served equally often before and after every product, and in all positions equally.

4. When using a balanced-block design, the complete permutation is served before repeating a serving order. This practice will minimize time-order effects.

5. Completely randomized designs are considerably less useful than balanced-block designs, as some serving orders will occur infrequently, whereas others will occur too frequently. This is not a great concern when the panel size is large enough to obtain 100 responses per product. However, when the tests have fifty or fewer responses per product, considerable order bias may occur.

6. If all products cannot be evaluated by each panelist in a single session, the order is balanced across the panel within each session. Balanced-block designs are preferred to balanced-incomplete-block designs, because they allow for all possible comparisons by each panelist and also use the fewest number of panelists. It is possible for each panelist to evaluate all products by conducting several sessions, thus avoiding the use of an incomplete-block design. Incomplete-block designs should be considered only after a decision is reached that the balanced-block design is not possible.

7. If it is not possible to have all panelists evaluate all products, serving order balance should be maintained across the panel.

8. A sequential, monadic serving order provides for greater control of the time interval between products, thereby controlling sensory interaction and providing for a more independent response to each product.

Selection of Consumer Panel

The selection of a consumer panel is discussed in greater detail in Chapter 4. Identification of the target consumer population for the product is crucial to the project. Accurate identification of the target population enables the project leader to design a sampling strategy that will give increased confidence in generalizing test results to the consumer target population. The *target population* may be defined as the segment the population that uses, or is expected to use, the product. The identification of this population is usually achieved through surveys or prior sales information, and so forth. The client can often provide this information. Once the panel characteristics are defined, recruitment and qualification of the panel follow. The number of panelists to recruit will be dependent on the experimental design. To obtain the required number of responses, it is desirable to overbook qualified participants by 20–50%. Experience of the project personnel will determine by how much to overbook to allow for "no-shows."

Test Schedule

The test schedule should be planned while keeping in mind certain characteristics of the consumer target market for the product. If the objectives call for involving a number of consumers who are employed during the day, some test sessions should be scheduled during the lunch hour and in the evening after the workday. School vacation schedules, holidays, and special events that may influence the ability of many consumers to show up for a test should likewise be considered.

When interviews with homemakers, many of whom have young children, are necessary, such as during home-use tests, these should be scheduled when interference from the children is least expected. Similarly, one should avoid time periods when the consumer is preparing supper.

Samples

The conditions of sample preparation and presentation should be identical to those under which a product is usually consumed. To prevent bias, it is essential that all samples for testing are selected, prepared, and served to the panel in a manner that will have limited impact on the perception of the products, and in a manner that is identical for all products.

Sample Screening

The samples tested should be representative. Initial testing of product acceptance is often conducted using samples prepared under laboratory scale conditions. After scale-up to commercial scale, samples produced during a regular production run may no longer be the same product that was originally tested. It is recommended that adequate testing of product prepared under regular production conditions be conducted. Furthermore, the product to be used in sensory tests should be held under conditions that simulate the normal distribution system.

Sample Variability

A certain amount of variability is present in products. However, the samples to be tested must be as consistent and uniform as possible with regard to production lot, age of the sample, package size, and similar factors. The objectives of the test should be considered in determining the sample preparation procedures to be used that would result in increased consistency and uniformity of the sample or help to determine the sources of variability in the sample. For example, in preparing frankfurters for a sensory test, opening ten packages, placing the contents in one container, and then randomly selecting ten pieces from the container for a sensory test would appear to provide the most consistent sample, but would prevent determination of product variability between packages.

Pretesting

Preliminary testing needs to be conducted whenever possible to enable the project leader to determine if the selected sample preparation and presentation procedures are appropriate for a specific test, to specify any special requirements related to preparation and serving, and to identify any additional procedures required as they relate to the product or test. The test objective will determine what test method to use, experimental design, test conditions, sample preparation, and serving procedures. The sample size, the number of samples to be evaluated, water for rinsing and palate cleansers, sample temperature, and lighting considerations should be considered. All testing conditions and sample preparation and serving conditions and procedures should be appropriately modified as a result of pretesting.

Sample Size

The test objectives are an important consideration when determining sample size. The sample size should be adequately large so that consumers will be able to evaluate product acceptance and any other attributes being evaluated. Sample weight and size should be uniform between samples and panelists. Standard scoops and measuring cups should be used to maintain consistency in sample volume, and top-loading balances should be used to standardize sample weight. Preliminary testing will help to determine the amounts of sample or number of pieces of sample to be placed in the mouth during evaluation. It is often necessary to provide instructions to panelists to ensure that enough products are evaluated to give a reliable result. For example, "Place at least two slices of carrots in your mouth" or "Drink at least half of the sample," before completing the evaluation. When sample size evaluated by consumers is critical, servers should monitor plate waste for the amount of sample remaining in the serving containers after evaluation.

Sample Preparation

Preparation procedures are dependent on the test objectives. In general, the samples must be consistent and uniform. Product characteristics should be considered when planning on sample preparation steps; Preparation methods should be clearly outlined. Preliminary testing is particularly useful during this stage of planning.

Preparation of all samples should be standardized such that all sources of variability due to the preparation and serving procedure should be eliminated, if not minimized, so that the only variability will be that inherent in the samples. When cooking samples, identical cooking units or appliances should be used (e.g., identical broilers for broiling hamburgers, identical microwave ovens for reheating frozen entrees). When samples cannot be prepared side-by-side in identical cooking units or appliances, cooking should be randomized among cooking units.

Many food samples usually require heating for a specified length of time to a specific endpoint temperature for microbiological safety and appropriate flavor development. When measuring cooking endpoints, monitoring with the use of an appropriate temperature-measuring device is necessary. The selection of the temperature-measuring device depends on the specific location in the sample where temperature needs to be monitored. For example, for roast beef or roast turkey, thermocouples with a temperature recording device may be appropriate for monitoring cooking and internal endpoint temperatures, but for hamburgers that need to be flipped during cooking, using a metal probe, specially designed to measure internal temperature of the patty, may be more appropriate to record internal temperature of the patty. The measuring device used should be consistent for all samples. Furthermore, their placement in the food where temperature has to be measured should be consistent throughout the food preparation process. Some food preparation procedures are described here.

Microwaving of Samples

Many products are heated or reheated using microwave ovens. If microwave ovens are used, their brand, model number, and wattage should be similar. As indicated earlier, it is advisable to randomize the ovens used in the preparation of samples to keep variability due to sample preparation at a minimum. It is recommended that the sample to be cooked or reheated in the microwave be similar in size, weight, and shape to ensure uniform sample heating rates. All ovens should be clean before use. Placement of the samples in the oven should be consistent. Uniform procedures for stirring and rotating samples will help to minimize hot or cold spots, as would use of turntables in each unit. Preliminary testing of the cooking procedures is a necessary step in determining final cooking or reheating procedures to be used. If several microwave ovens will be operating at once, they must be tested and calibrated while all are operating to ensure that circuit voltages are adequate and supply uniform power to the appliances.

Baking or Broiling

Preliminary testing the baking or broiling cooking procedures is necessary in determining final cooking or reheating procedures to be used for the evaluations. It will be necessary to standardize and monitor oven temperature prior to and throughout the test. Do not combine the use of gas and electric ovens in one test. Containers used for baking or roasting should be appropriate in size and shape, and uniform between samples. Be consistent in placing samples in the same position in the oven. Standardize the checking of samples to keep opening and closing the oven door at a minimum. Identical cooking time may not result in the same internal endpoint temperature; therefore, temperature is preferred to time as the indicator of doneness.

Stovetop Cooking

Electric and gas burner units heat at different rates; therefore, only gas or electric-heating units should be used during sample preparation. Use identical electric heating element size or flame size in the case of gas. When heating food, cooking should be done in the same type and size of container. When water or oil is necessary to cook the samples, identical amounts of sample should be cooked in identical amounts of water or oil. In cases wherein there are three or more samples, there may not be sufficient cooking units of the same size and wattage; therefore, randomization between cooking units is appropriate. Control of hot spots or any other sources of variability is important. When temperature maintenance is necessary, consider the use of stainless steel double boilers.

Grilling or Pan Frying

It is recommended that the same model and brand of electric skillets or grills be used in the preparation of samples. If several electric skillets or grills will be operated at once, they must be tested and calibrated while all are operating to ensure that circuit voltages are adequate and supply uniform power to the appliances. The temperature controls of these appliances should be calibrated and temperatures monitored during cooking. Surface temperature of food-service grills should likewise be monitored. Be consistent in the placement of products on these surfaces and avoid cooking samples in either hot or cold zones. If oil or a noncaloric frying spray is used to ensure that the product does not stick to cooking surfaces, its use should be consistent from sample to sample. For food-service grills, it may be necessary to clean and season the grill surface between samples. Procedures for such steps during sample preparation should be outlined in the cooking procedures.

Deep Frying

Fill the deep fryer to the recommended fill line for oil. It is desirable to maintain this level throughout the test to keep the product immersed while cooking yet prevent the oil from overflowing. Determine recommended frying temperatures for the food to be fried and conduct preliminary testing of the frying procedure to determine the cooking endpoint. During the preliminary testing and the actual test, allow sufficient time for the oil to reach the desired frying temperature. Monitor the temperature of the cooking oil during frying. During the test date, it is desirable to fry at least one or two batches of product prior to frying actual samples that will be evaluated. Exercise caution when frying food that has a tendency to spatter. If possible, fry under a hood to prevent frying odors from dissipating. Finally, and most importantly, exercise extreme care to prevent fires from occurring. Know the location of the nearest fire extinguisher, and learn how to use it prior to the pretest date.

Minimize the holding time between removal of food from the fryer and serving of samples. Preliminary testing will help to determine maximum holding time for the samples without affecting the sensory properties of the samples. Preliminary testing will likewise determine whether heating lamps or other devices to maintain the samples at the desired temperature should be used. A constant holding time between frying and serving of the samples should be maintained.

Sample Presentation

Serving Temperature

The serving temperature of a sample may influence its moisture content, texture, viscosity, and other characteristics important in determining consumer acceptance. Sample serving temperature is determined by the objectives of the test. The serving temperature must be consistent between samples and within a predetermined range. For example, the temperature of precooked and reheated food needs to be maintained over 60°C (140°F) for microbiological safety. Reheating of a food may result in changes in flavor, texture, and appearance of the samples and is not recommended unless the objectives of the test require it. Preheating serving containers, such as small glass petri dishes for ground beef samples, will help to minimize heat loss between serving and evaluation when use of a warming device is not recommended. If holding the food will not appreciably affect it, the selected holding temperature range may be maintained using steam tables or water baths, heating lamps, hot plates or tray warmers, heated sand, crushed ice, or ice-water baths. When these are used, care should be exercised to make certain that devices or equipment should be identical in terms of wattage, source of energy, height of lamps, and amount and depth of water in steam trays or water baths. Sample temperatures of hot food should be such that the food can be held comfortably in the mouth. The temperature of the samples should be monitored throughout serving and evaluation. It may be necessary to provide consumer panelists with a new sample during a test period to maintain a consistent temperature.

When cooking results in variability within a sample, such as browner surface color compared to the internal color in cakes or roasts, a decision must be made prior to the test date on what portions of the cooked product will be served to panelists. Whether an internal sample or the whole product is served, care must be exercised to ensure uniformity throughout and that panelist will receive similar portions to evaluate.

Sample Holding Time

Samples often change between preparation and serving, and it is critical to control the time between sample preparation and serving. For example, beef steaks or hamburgers cool down considerably after cooking, and snack foods may absorb moisture from the atmosphere and become limp. To prevent this, it may be necessary to schedule the preparation of samples such as hamburgers to be cooked and served with as short a holding time as possible. On the other hand, if the test objective is to determine the acceptability of fried potatoes held under heating lamps for various time periods, scheduling should allow for the required holding period between preparation and serving.

General Guidelines for Preparation of Specific Foods

Beverages

Different types of beverages have different sample-preparation requirements. When preparing beverages or drinks from a powder, these are usually reconstituted with either water or milk. If reconstituting with water, use triple-distilled water or deionized water, which is readily available in many food testing or analytical laboratories, or use bottle water. When reconstituting with milk, determine what fat level the milk should possess and use this type of milk for all samples. Milk can be purchased from a local dairy in five-gallon bag-in-box containers. When using milk for smaller sized containers such as gallon jugs or cartons, combine the contents from different containers into one uniform lot to minimize variability due to the milk used and provide consistency. In some cases, powdered beverages need to be blended after reconstitution to provide smoothness of texture.

Beverages or drinks that are blended from one or more liquids with water or milk should be served immediately after preparation. Carbonated beverages should be poured and immediately served to the panel. When larger sized containers are used, beverages should be held for only as long as sufficient carbonation is maintained. Preliminary testing would help to determine the holding time for larger sized containers of carbonated beverages before the beverage should be discarded. Juice products such as those held in boxes or cans should be shaken prior to pouring, then immediately served to the panel.

Hot beverages such as coffee or tea should be served from appropriate serving containers such as insulated flasks that are easy to serve from and maintain serving temperature. Flasks of similar capacity should be used for each sample.

Cold beverages, such as carbonated beverages, chocolate drinks, and milk, are best served at approximately 5–9°C (42–48°F), hot beverages such as cocoa and hot chocolate at 60–66°C (140–150°F), and hot coffee and tea at 66–71°C (150–160°F).

Baked Goods

This product category includes breads, quick breads, biscuits and muffins, cookies, crackers, and cakes. One important consideration to minimize variability between samples is to serve samples from the center part of a loaf of bread or a slice of cake and exclude the crusted ends or browned edges, respectively, unless the test objectives requires it. When appearance is to be evaluated, the sample size should be large enough to allow the panelists to evaluate appearance. For example, a whole slice of bread, a muffin, or a roll would be required in order to evaluate their appearance. Baked goods should be served right side up.

Overbrowned samples, such as overdone cookies and broken pieces of product, should be not been served. Samples such as sliced bread and cake may be held in sealed plastic bags when setting up trays and during the test to prevent drying out of samples. Most baked goods may be served at ambient temperature.

Cereals

Minimal preparation is required when testing dry or ready-to-eat cereals. Dry cereal is usually served with milk and often with sugar, which the panelists themselves add to the cereal. The maximum holding time for cereals should be determined through preliminary testing. Dry cereals and ready-to-eat cereals may be served at ambient temperature, whereas hot cereal should be served at 43–49°C (110–120°F).

Candy

Candy is served at ambient temperature. Large candy bars need to be cut into smaller pieces approximately 2.5 cm (1 in) long if larger pieces are recognizable by their appearance or any brand identifiers. For smaller sized candies, serving approximately three pieces is sufficient. It may be necessary to specify to consumer panelists how many samples they should consume when making evaluations.

Canned fruits

Canned fruits such as applesauce may be served at either ambient or refrigeration temperatures as the objectives require.

Casseroles

Casseroles should be served hot, at 60–66°C (140–150°F). Each sample should contain each casserole component.

Cheese

Trim the outer portion of block cheeses before samples. Waxed surfaces and wrappers should be removed before serving. Samples should not be allowed to dry out and may be wrapped with plastic after slicing to prevent loss of moisture and then be returned to the refrigerator. Samples may be cut to 2.5 or 1.25 cm cubes. A cheese wire may be useful for cutting soft cheeses and may be wrapped with plastic after slicing to prevent loss of moisture and then be returned to the refrigerator. Sliced cheese may be served with white bread or one-fourth of a sandwich. Although cheese should be stored in the refrigerator, cheese should be removed to allow enough time for the cheese to equilibrate twenty to thirty minutes before serving and be served at ambient temperature. The plastic wrap may be removed just before serving.

Coffee Whitener

Coffee whitener products should be served with coffee at a serving temperature of 66–71°C (150–160°F).

Cottage Cheese, Sour Cream, and Yogurt

The serving temperature for cottage cheese, sour cream, and yogurt is either ambient or 4–10°C (40–50°F) depending on the test objective. Decide, along with the client, prior to the test date whether any liquid that has separated due to syneresis will be reincorporated into the samples for evaluation. Scoops may be used to dispense the samples. When scoops are used, the scooping procedures should ensure sample uniformity. The objectives of the study will determine if yogurt should be stirred prior to evaluation, especially if products have fruit at the bottom. A carrier such as baked potatoes or potato chips may be used for sour cream products.

Fat and Oils

When testing oil products for cooking or frying foods such as doughnuts or french fries, the food should be served hot from the grill or fryer. Margarine, butter, or spreads must be served at ambient or refrigerated temperature depending on the objectives of the study. Sliced, firm-textured white bread or unsalted crackers may be used as carriers. A nonserrated metal knife is ideal for spreading and should be provided to panelists.

Jams, Jellies, Preserves, or Spreads

These are best served at ambient temperature using slices of a firm-textured white bread or unsalted crackers as carriers. A metal nonserrated knife to be used for spreading the sample on the carrier should be supplied.

Fish

Fish products should be served at 66°C (150°F). Fish products are often fried or baked from the frozen state.

Meats

Samples should be prepared identically. They should be similar in shape and physical condition. For example, the muscles used should be identical; the direction of the grain of the meat should be uniform. Larger samples should be cut to uniform size. The use of a template to cut the meat to the desired size is recommended. If cutting samples to a uniform size, odd-shaped pieces should be discarded. The cooking methods used should be identical for all samples.

Beef patties should be cooked to 71°C (160°F) for microbiological safety. Beef stew is best served at 68°C (155°F). Bacon, frankfurters, and sausages may be served at ambient temperatures or 60–66°C (140–150°F). Ham maybe served at 10°C (50°F) or hot at 60–66°C (140–150°F). Other meats such as pork chops or roast beef steaks, or roasts, sausage links, or patties should be cooked to 63°C (145°F) and served at 60–66°C (140–150°F). Luncheon meats, canned chop meats, and cold meats such as salami and bologna should be refrigerated until served between 2 and 5°C (35 and 40°F).

Milk

Milk is commonly evaluated at 12.8°C–18.3°C (55–65°F). The serving size depends on the objectives of the study. However, 30 ml to 120 ml (1–4 ounces) in glasses is appropriate for most tests.

Pancakes and Waffles

If prepared from dry mixes, these should be uniformly prepared. A mixer should be used to ensure uniform mixing. Preliminary trials should be conducted to determine how long the batter needs to stand, if at all, before pouring. A standard scoop should be used to dispense the batter for portion control. If a seasoned griddle is used, oil will not be needed. If oil is used to season the griddle, the procedure should be uniform throughout.

Poultry

Poultry breast meat should be cooked to 63°C (145°F) and served at 60°C (140°F).

Peanut Butter

Peanut butter should be served at ambient temperature. Sliced, firm-textured white bread may be used as a carrier; a nonserrated metal knife is ideal for spreading and should be provided to panelists.

Refrigerated and Frozen Foods

Refrigerated foods should be held below 4°C (40°F) for microbiological safety. Frozen foods should be maintained frozen solid for maximum quality until cooked or served. Furthermore, samples should be adequately protected from drying out or from absorbing odors from other foods in the refrigerated or frozen space during storage. Temperature of refrigerators or cold rooms where the samples are to be stored or held should be monitored closely prior to use. When loading refrigerated space with samples, sufficient airflow around sample packages should be maintained. If the samples are to be held for extended periods as in shelf studies, the cycling patterns and the temperature fluctuations in the refrigerated or frozen storage space should be monitored prior to and during the storage of samples. When samples need to held for short periods of time prior to the test where a refrigerator is not available, such as in central location tests, refrigerated foods maybe held in ice chests or coolers, insulated containers, water or ice baths, or in other devices that help to maintain the temperature of refrigerated food. In some instances, the use of dry ice is necessary to maintain the temperature of frozen foods. When using dry ice, samples should be sufficiently protected from direct contact with the dry ice. Dry ice should be handled with caution to avoid burns and be used only in well-ventilated rooms.

The temperatures for storing, preparing, and serving refrigerated and frozen foods should be based on the test objectives. Preliminary testing and temperature monitoring should be conducted to determine proper handling protocol for samples. Frozen entrees are usually served at 60–71°C (140–160°F). Frozen desserts are best served at −18 to −10°C (0–14°F). Samples may be scooped or sliced. Scooping procedures should ensure uniformity of the samples.

Pies

Pies may be served at refrigeration, ambient, or warm temperature as appropriate. They should be served with the crust unless the test requires that only the filling should be tested.

Salad Dressings

Preparation of the salad dressings must be identical. The samples may be served at ambient or refrigerated temperature, but temperature must be consistent between samples. Carriers such as chopped, well-dried iceberg or precut lettuce may be used. In such cases, a specific amount of salad dressing may be poured on a predetermined amount of cut lettuce and tossed using the same number of strokes for each sample.

Sauces and Gravies

Sauces and gravies may be served at 60–71°C (140–160°F). Instant mashed potatoes may be used as a carrier for gravy. Unsalted tortilla chips may be used for cheese sauce. Unsalted spaghetti may be a carrier for spaghetti sauces and plain boiled white rice for strong flavored sauces such as soy sauce.

Snack Foods

Snack foods such as corn, tortilla, and potato chips should be served at ambient temperature. These should be served from freshly opened packages or airtight containers.

Soups

Soups should be served hot at 60–71°C (140–160°F). Each sample should contain each of the soup's components.

Syrups

Syrups are most often served at ambient temperature. These may be served with plain pancakes or waffles reheated from the frozen state.

Sample Presentation

This should be planned to prevent bias. Environmental conditions to consider include special lighting needs, if needed, and other instructions as needed by the requester or client.

Sample Containers

Similar type of containers and utensils should be used throughout the test. In selecting a container or utensil to be used in the test, the following factors should be considered:

1. The test objective
2. Test-material characteristics
3. Interaction between test materials and sample container
4. Volume of sample required by test method
5. Ease of handling the product during evaluation
6. Characteristics of the container

The test objective should be considered in determining the container to be used. For example, when measuring acceptance of beverages such as flavored milk and juice, these should preferably be served in clear drinking glasses rather than in opaque soufflé cups with lids. Test-material characteristics such as the sample size and shape, physical state (solid or liquid), or serving temperature are among the most important considerations, as are any interactions between the test material and the sample container. The volume of the sample needed to complete evaluation is critical and would influence choice of the container. For example, consumption volume has been used as a measure of snack-food acceptance. In such cases, a bag would be a more appropriate container for the snack food than a small plate. Ease of handling the product during evaluation should be considered when determining serving utensils. For example, when testing acceptance of marinated steaks from different grades of beef, stainless steel forks and knives would be more appropriate to use than plastic utensils that may easily break. Finally, the characteristics of the sample container should be considered. These should preferably be neutral in color unless specific reasons exist for using tinted containers, and they should be made from materials that are odor-free. If glassware is used, it should be washed with unscented detergent and rinsed thoroughly prior to use. Others recommend baking glassware at 93°C (200°F) for several hours to expel sources of odors.

Palate Cleansers

Palate cleansers are used before evaluation and in between samples. The type of palate cleanser that can be used is dependent on the type of sample being evaluated. A palate cleanser needs to be bland and easily cleared from the mouth. Bottled, filtered, or distilled water is an effective cleanser. Unsalted saltine crackers or water biscuits have been used effectively. Warm water has been effectively used in evaluating certain products such as vegetable oils, but these have been descriptive rather than affective tests.

Carriers

Carriers are most often used to present products in a manner that is more typical of how the product is used by consumers. Some products need to be evaluated on a carrier. The use and selection of the carrier depends on the objectives of the test and appropriateness for the product. A carrier can be considered appropriate when use of the carrier

1. Represents traditional food-use patterns
2. Has a neutral influence on the evaluation of the product
3. Does not mask product attributes
4. Does not distract attention from the product

Traditional food-use patterns often justify the use of a carrier. Examples are the use of milk when testing a dry breakfast cereal and a slice of white, firm-textured bread to evaluate spreadability of a peanut butter. The carrier should have a neutral influence on the evaluation of the product, and should elicit neither positive nor negative comments. For example, in testing strong-flavored sauce product such as soy sauce, white rice would be a good choice for a carrier. The carrier should not mask product attributes; therefore, carriers should preferably be less strong flavored than the test sample. Finally, the carrier should not distract attention from the product. For example, when a sugarless syrup-substitute product is tested, plain pancakes or waffles may be used without distracting attention from the product.

When using a carrier, it is important that the use of the carrier be consistent throughout the evaluation. The sensory properties of the carrier—appearance, aroma, flavor, and texture—and the amount and temperature of the carrier should be kept constant.

Sample Number

Usually, only one or two samples are involved in a consumer test—the control, or reference, product and a test sample. When a single sample is presented, this is referred to as a monadic presentation. Samples may also be served simultaneously or sequentially.

It is necessary to limit the number of samples to a number that can be handled by the panelists. The decision on the number of samples in a test session should consider sample characteristics such as flavor, the number of questions to be asked, and panelist fatigue. Increasing the number of samples presented to each panelist may increase confusion (ASTM, 1979) or cause boredom or fatigue. Panelist fatigue (physiological and psychological) should be considered when determining the number of samples to be evaluated in one session. In some cases, prescreening may be used to limit the number of samples to be tested. Although it is ideal to have the entire panel evaluate all samples, it may be not be possible in some cases. When it is necessary to test several samples in a consumer test and panel members are given only a few samples to evaluate, the experimental design becomes more complicated and more panelists are needed.

Number of Responses per Sample

The number of responses per sample is influenced by the specific sensory characteristics of the product, how confident one is with the subject selection, the type of information desired, and the experimental design (ASTM, 1979). Usually, a consumer test will involve not less than fifty responses per product, and the number depends on test objectives and increases with the importance of the decision derived from the results (possibly 400 to 500). In a consumer test, there is a preference for a larger panel size. However, the total number of responses is also significantly influenced by the availability and cost for each consumer participant. Obtaining more than one response per sample from each subject can be valuable. It allows for a measurement of variability within a subject as well as across subjects.

Sample Coding Procedures

A sample coding scheme should attempt to eliminate the likelihood that the subject response will be influenced by the sample code rather than the product.

Care must be taken to use sample codes that are not meaningful to panelists and that do not give the subject any clues as to the product's identity. Single-number or letter codes may influence panelists who may have biases toward their initials or certain numbers (such as 1 or 7) or two-digit code numbers (11, 21). Two systems for sample coding have been found useful. The more widely used practice is to use a three-digit code from a random number table or generate the random numbers electronically. The second is a single-letter multiple coding system in which an entire series of letters (six or eight) will be used for one sample, one letter at a time, and another series of six or eight letters will be used for the other sample, one letter at a time (ASTM, 1979).

In certain cases, it may not be practical to use a multiple coding system. An example of such a case is when using a large population sample in a mail-out home-use study. If the study is a paired-comparison test, a convenient solution proposed by ASTM (1979) is to split the population in half. Use a single, neutral letter to represent one test sample, and another single, neutral letter to represent the other test sample. Send the samples out to one-half of the sample population and then reverse the codes before sending the samples out to the second half of the population. Thus, if Sample 1 is coded K and the other sample, Sample 2, is coded M for one-half of the population, the second half of the sample would receive Sample 1 labeled M and Sample 2 labeled K.

When coding samples, grease pencils or coded stick-on labels are recommended. Felt pens or markers may be used as long as they do not impart an odor. In some instances, the odor from such pens disappears with time; therefore, if used, sample cups should be marked far enough in advance to allow the odor to disappear.

Product Evaluation and Data Collection

Test Controls

Time Intervals between Samples

A preliminary testing will determine how much time to allow each panelist to evaluate each sample. Sufficient time between samples and a set of samples should be allowed for each panelist to evaluate each sample and recover or be equilibrated between samples. If a specific time interval is required between samples and a set of samples, it is necessary to ensure that all panelists maintain the time intervals. Maintenance of the appropriate time interval between panelists is easily achieved when using computerized sensory evaluation programs with the capability for a specific time interval to be programmed between samples. In such cases, a message on the video display should indicate the time interval before the next sample will be served and whether the panelists should sit in their booths and wait for the next sample or leave the booth and return after a specific time interval. If computers are not used, the servers can control the time intervals by serving each sample only after the required time interval is reached.

Expectoration

Differences in expectoration may affect the results of the test. It is therefore necessary to carefully consider the objectives of the test to determine whether panelists need to expectorate. When expectoration is an option in a consumer affective test, individual panelists need to be consistent from sample to sample.

If panelists will expectorate, appropriate and sanitary containers should be provided for this purpose. Usually, 16-ounce disposable, opaque cups with lids are provided for this purpose. The opaque cups allow discreet expectoration of samples. The small number of samples usually recommended in a consumer test will ordinarily not require the use of more than one cup for expectoration; however, extra cups should be available for this purpose. Panelists should discard their own containers.

Test Facility and Environment

The test facility should be scheduled with enough lead time before the test date. Scheduling is important whether the test will be conducted in the company laboratory or in another location. The testing environment should be quiet and odor free. Communication between panelists should be prevented. Conversation between panelists and sample servers should be kept to a minimum and, if necessary, should be as quiet as possible. Other external influences that may affect panelists' responses should be eliminated or at least minimized. The testing environment is more readily controlled when conducting consumer affective tests in a sensory laboratory, and less readily controlled in a central location. The least control over the testing environment is experienced when conducting a home-use test.

Written Instructions and Briefings

It is usually necessary during the planning stages for the project leader to provide laboratory or field-workers with a complete set of written instructions covering the test, a copy of the final questionnaire that has been pretested and revised accordingly, and all other items necessary to ensure that the test will run smoothly. In a briefing session before the actual test, it is essential that the project leader review the complete test protocol to determine whether the test personnel, in the laboratory or field, are familiar with their assigned tasks, the procedures, and the serving instructions. In addition to the briefing session, it is recommended that a complete dry run of all testing procedures be conducted on the test date, from the preparation of samples and orientation of panelists to actual test procedures to be conducted using two or three untrained individuals as panelists.

The Interview

The interviewers are usually briefed and trained in interviewing procedures by the field supervisor or project leader. It is advisable for the project leader, the field supervisor, and all interviewers to be present at a "dry run" wherein the entire process of interviewing is conducted using individuals who assume the role of consumers in trial interviews. Control of the testing procedures should ensure that each interview is consistent and that the respondents provide useful information. The project leader needs to be able to determine whether the information being gathered will be usable. It is the interviewer's responsibility therefore to provide feedback by reporting any difficulties or deviation from expectations to the field supervisor immediately. The interviewer should continue to observe each respondent during the course of the interview and judge the ability of each respondent to understand and complete each part of the questionnaire. In cases of poor performance, the interview should be discontinued.

Start and Completion Dates

A schedule is necessary for program and budget management. Adherence to a schedule is important to minimize project overruns. In addition, the schedule helps to identify potential problems, including conflict with other test schedules. The schedule provides the team and the client with precise information on the proposed starting and completion date. For project personnel, the schedule should contain, in addition to the starting and completion dates, the decision points, individuals responsible for these actions, and the names of report recipients (ASTM, 1979). This schedule should be brief and incorporated into the project plan.

Duplication of Forms

Ballots, informed consent forms, and receipt of incentive or honorarium forms need to be duplicated at this stage. If the test will involve computerized ballots, the ballot and all hardware such as video terminals and light pens that will be used during the test should be tested. Instructions to panelists that need to be posted in the booth area should also be duplicated.

Rewards and Incentives

Rewards and incentives to panelists are important considerations during the planning of a consumer test. The amount of the incentive paid to participants may differ according to such factors as the distance traveled by the panelist to the test location, the incidence rate of qualified participants, the length of the session, and the difficulty of the tasks involved.

The incentive is usually given after the session is completed. Incentives for participation may be in the form of cash, selection from a gift catalogue, gift certificates, and tickets to special functions such as ball games or concerts, or donation to charity or a nonprofit organization.

Data Analysis and Processing

The methods for data analysis and processing should be outlined at the start of the experiment. If a statistician or data-processing specialist is used, consultation with this individual during the initial stages of project planning should occur. Consultation should be conducted as early in the project planning and implementation as possible to increase efficiency of data processing and preparation of figures and tables.

References

Amerine, M. A., Pangborn, R. M., and Roessler, E. B. 1965. *Principles of Sensory Evaluation of Food*. Academic Press, New York.

ASTM, Committee E-18. 1979. *ASTM Manual on Consumer Sensory Evaluation*, ASTM Special Technical Publication 682, E. E. Schaefer, ed. American Society for Testing and Materials, Philadelphia, PA, pp. 28–30.

Box, G. F., Hunter, W. G., and Hunter, J. S. 1978. *Statistics for Experimenters*. Wiley, New York.

Cochran, W. G., and Cox, G. M. 1957. *Experimental Design*, 2nd ed. Wiley, New York.

Jones, L. V., Peryam, D. R., and Thurstone, L. L. 1955. Development of a scale for measuring soldiers' food preference. *Food Res.* 20:512–520.

Lawless, H. T., and Heymann, H. 1997. Sensory Evaluation of Food: Principles and Practices. Chapman & Hall, New York.

Peryam, D. R., and Pilgrim, F. J. 1957. Hedonic scale method of measuring food preference. *Food Technol.* 11(9):9–14.

Stone, H., and Sidel, J. L. 1985. *Sensory Evaluation Practices*. Academic Press, San Diego, CA.

Stone, H., and Sidel, J. L. 1993. *Sensory Evaluation Practices*, 2nd ed. Academic Press, San Diego, CA.

4

The Consumer Panel

Introduction

The consumer panel is one of the most important critical factors to consider when conducting a consumer affective test. In a consumer test, a sample of consumers is selected to find out whether they like a food product sample. In conducting the tests, we assume that our measurements, which are made using the sample of consumers, are more or less representative of the entire population of people who would purchase the product. For consumer testing, the panel must be representative of the target market for the product, so that panelists do not give misleading information about how consumers will like the product. The target market consists of those consumers who would actually purchase and use the product. Statistical tests require that a sample be selected randomly from the population it is representing (O'Mahony, 1986). If the panel does not represent the target market, the data will have little or no predictive value.

Sampling

Sampling is one of the most important considerations in conducting consumer sensory research. Sampling allows consumer testing using a smaller number of panelists as opposed to testing of the entire target population. Sampling involves statistics and judgment. Furthermore, the process of sampling requires a decision to be made regarding the balance between the usefulness of the information gathered and the return on the investment of obtaining this information. In other words, consumer researchers need to balance the need to identify and use a sample of consumers who represent the target population against the cost of having a precise demographic model. If the sample is drawn so that it is representative of a population, the conclusions form the statistical analysis apply only to that population.

In the conduct of many consumer affective tests, true random selection of the consumer panel was rarely achieved (Stone and Sidel, 1993). Although attempts should be made to recruit a consumer panel that truly represents the target market for the product, many tests usually involve a limited number of participants from a larger group of prescreened and qualified consumer, often from an existing database of consumers. The more extensive databases allow for recruitment of a panel that better represents the target market. Ultimately, selection of the consumer panel for a specific test will depend on many factors and the test objectives.

Calculations for Sample Size

One of the features of the sensory acceptance test that is most often cited as a weakness is the relatively small number of consumers compared with the large numbers in a market test, which has usually 100 to 500 or more respondents. This belief originates from the misconception that validity is directly associated with the use of a large number of consumers.

A larger panel of consumers in a test enhances the likelihood of finding product differences and of determining which product is preferred (Stone and Sidel, 1993). However, other factors, such as the product, and practical and statistical considerations, must be taken into account before one mindlessly increases the number of products in a test. The problem of time and cost of doing the research prevent the collection of acceptance data from each potential consumer of the product. The application of statistical sampling techniques that permit one to use a subset of the population allows one to generalize the results.

There are several methods that have been employed to derive the sample size to be used. In some cases, sample size is selected using no basis except intuition. This process is inefficient and often leads to over- or undersampling. Oversampling will mean wasted time and resources, whereas undersampling may lead to the need to conduct another test. Usually a 95% confidence that a difference exists between two products, or that the product will be rated at a specified level by the entire population, is desired (ASTM, 1979).

Calculations of sample size will depend on three factors: (1) the risk that management is willing to take, (2) the standard deviation from previous tests, and (3) the difference in quality of the samples that seem to be important to users (ASTM, 1979). Most statistical texts will give a formula for calculating sample size.

Sample size is a concern to management or to clients who want to maximize the success of their decisions but also want to save time and resources in conducting the test. Management sets the degree of precision in which it would be interested. This tolerance level may be "within 5%" of the figure for the total population. At the same time, management sets a confidence level at a desired value, for example, an error of 5% (ASTM, 1979).

Table 4.1   Factors That Influence the Sampling Plan

---

1. The business decision that will be based on information from the sample.
2. The practicality of measuring a portion of the universe.
3. Costs of sampling, the entire project and the return on information investment.
4. The level of risk that management is willing to take, such as Type I and Type II error (wherein Type I error consists of rejecting the null hypothesis when it is true, and Type II error consists of accepting the null hypothesis when the test hypothesis is true) and other risk factors.
5. Demographic characteristics of the population.

The sampling error is the difference between the measurements from the sample and measurements from the universe of consumers that make up the target population, if similar techniques have been used to collect the data. Precision is the degree to which the sample estimate approximates the value of a 100% count. The degree to which the estimate approximates the true figure of the population for the universe is referred to as accuracy (ASTM, 1979).

The variability of results from the sampling plan can be affected by bias and sampling error. Bias is the difference between the true value of the sample and the measure value derived from the sample. In preparing a sampling plan, several factors need to be considered. These are listed in Table 4.1.

Most sampling approaches used in sensory evaluation are researcher controlled and involve either quota or chunk (convenience) sampling approaches. Quota sampling is one in which a definite number of individuals belonging to a possibly relevant stratum are specified for inclusion. Relevant strata would be age, sex, income, and product usage, among others. If a population has 50% males, the sample will include that many males. If the population has 34% of consumers above the age of 55, so will the sample.

Chunk or convenience sampling uses those respondents who are most convenient. Judgment sampling is one chosen to reflect what one believes the market to be. It can be a probability quota or chunk sampling that is modified to exclude individuals who do not fit predetermined judgments of a market. Sampling approaches relevant to the different consumer testing method will be covered under the different chapters devoted to each method. However, in laboratory tests involving consumers, chunk or judgment samplings are involved. In focus groups, judgment samplings are recommended. In central location tests, samplings may be either chunk or judgment. In laboratory tests involving employees, a chunk sampling is often used, but judgment sampling can be used. In central location tests, chunk or judgment samplings are used.

Panel Size for Various Consumer Tests

Focus Groups

Focus groups vary in size and ideally consist of 8 to 12 participants who represent the projected target market (ASTM, 1979; Sokolow, 1988; Chambers and Smith, 1991). The small panel size makes it difficult to draw a focus group sample that is truly representative of typical consumers or anticipated consumers of the product type involved. A panel of 10 is ideal; a panel with fewer than 8 participants does not provide adequate input, whereas one with greater than 10 participants will be likely not to provide sufficient time during a one-and-a-half to two-hour session to cover the number of issues desired.

Laboratory Tests

The panel consists of consumers who are recruited and screened for eligibility to participate in the tests from a consumer database consisting of prerecruited consumers. Usually, twenty-five to fifty responses are obtained; at least forty responses per product are recommended by Stone and Sidel (1993); however, fifty to 100 responses are considered desirable (IFT/SED, 1981). In a test consisting of 24 panelists, it may be difficult to establish a statistically significant difference in a test with the small number of panelists. However, it is still possible to identify trends and to provide direction to the requestor. With 50–100 responses per product, statistical significance increases to a large extent.

Sidel and Stone (1976) provided a guide for selecting panel size based on the expected degree of difference in ratings from the 9-point hedonic scale and the size of the standard deviations. They warned their readers that different products have their own requirements, indicating that smaller panels can provide statistical significance when the variability of the sample is small, and larger panels are needed as variability of the products increase.

not very well explained

Central Location Tests

Usually 100 (Stone and Sidel, 1993) or more consumers (responses per product) are obtained, but the number may range from 50 to 300 (Meilgaard et al., 1991), especially when segmentation is anticipated (Stone and Sidel, 1993). The increase in the number of consumers in the central location test compared to the laboratory test is necessary to counterbalance the expected increase in variability due to the inexperience of the consumer participants and the "novelty of the situation" (Stone and Sidel, 1993). Several central locations may be used in the evaluation of a product. Tests that use "real" consumers have considerable face validity and credibility. The increased number of respondents of 100 or more has advantages and disadvantages compared to the laboratory test.

wow!

Home-Use Tests

Due to the lack of control over conditions of testing in a home-use test, a larger sample than that required for a laboratory test is recommended. The minimum number of responses is usually 50–100 per product. This number varies with the type of product being tested and with the experience of the respondent in participating in home-use tests. With "test-wise" panelists, Stone and Sidel (1993) recommend that a reduction in panel size could be made, because these panelists are familiar with how the test is conducted and feel more comfortable in the test situation. In home-use tests that are multicity tests, 75–300 responses are obtained per city in three or four cities (Meilgaard et al., 1991).

Panel Selection

Whenever a consumer sensory test is conducted, a group of consumers is selected as a sample of some larger population about which the sensory analyst hopes to draw some conclusions. Therefore, it is important that the panel should match the target population. Usually, the sample is drawn from the target population.

Identification of the target population is essential to the success of the study. Proper identification enables the project leader to recruit an appropriate sample and to increase confidence in generalizing the results to the target population.

The *target population* may be defined as that segment of the population that regularly purchases, or is expected to use, the product or the product category. The panelists in a consumer sensory test should be selected on the basis of their demographic characteristics, frequency of use of a product, or product preferences. For example, children who drink milk should be used in testing different varieties of flavored milk for children; they do not need to drink a specific brand of milk unless the objectives of the project calls for this. For a presweetened breakfast cereal, the target population may be children between the ages of four and twelve (Meilgaard et al., 1991).

A screening questionnaire for product usage must be constructed by the project leader. An example of a screening questionnaire is shown in Table 4.2. The desired frequency of use must be discussed with the client or requester. For example, a client may choose to use a panel that has 60% heavy users and 40% regular users of a food product category. The screener may be a separate questionnaire administered during recruitment, whether by telephone or face to face. In addition, a few questions may be added to the questionnaire to verify responses.

Table 4.2   Sample of a Screening Questionnaire for Panelists Who Are in a Database

---

Prescreen panelist using database information prior to calling.

Sample quotas for this study are:

Age	Male	Female
18–24	6	6
25–34	6	6
35–44	6	6
45–54	6	6
55–64	6	6

Panelist Code #_____

Date Called and Status of Call:

Hello, my name is _____ and I am calling from *The University of Georgia Experiment Station from the Department of Food Science.* We are interested in consumers' opinions about *peanut products.* Do you have a few minutes to answer some questions about yourself and your use of peanut products? (YES—CONTINUE to 1; NO—TERMINATE and note the reason above under status of call)

1. What year were you born? (Write year _____ and place a check in the consumer's age group)
   18–34 _____ (1962–1978)
   35–54 _____ (1946–1961)
   55–65 _____ (1931–1945)
      (If over under 18 or above 65, TERMINATE. State "We are only supposed to survey those between the ages of 18 and 65, but I thank you for your time, and hope you will participate in other surveys that we may be calling for in the future. I hope you have a nice day.)

2. When was the last time you participated in any test with us?

   _____
         (If less than 3 months, TERMINATE)
      If answer is "I do not remember," prompt consumer with foods such as 'was it akara, honey spread, steak, hamburger, (if answer is "YES" to any of these, TERMINATE because consumer participated in a test within last 6 months). Otherwise, CONTINUE.

3. Do you have any food allergies?
   Yes _____
   No _____ Skip to Q#4
      If "YES," ask, "Which foods are you allergic to?" _____
   (If none of the foods mentioned include peanuts, CONTINUE. If allergic to peanuts, TERMINATE).

4. How often do you eat peanuts?
   If once a month or more, _____ CONTINUE schedule for test
   If less, _____ TERMINATE.

continued

Table 4.2 *Continued.*

5. We are conducting a taste test using some of the foods that I mentioned before. The test is scheduled for October 29 and 30. It will take approximately 1 hour and 20 minutes, which includes a 20 minute break, and you will be paid $10 for your time. Would you be interested in participating?

   (If "YES," CONTINUE) What would be the most convenient time and day for you to come for the test? (Check available times and dates)

   Which day would be best?
   Tuesday, October 29   _____
   Wednesday, October 30   _____

   The times for the test each day are:
              10:00 A.M.   _____
              2:30 P.M.   _____
              5:30 P.M.   _____

       If "YES," schedule for test. We can schedule up to 18 consumers per test but no more than a total of 60 consumers for the entire test.

       If none of the times are acceptable, TERMINATE. State, "I am sorry none of the times fit into your schedule. May I put you on a waiting list for this test, and you can call us if you find out you can participate in one of the sessions?"

6. We need to verify your name, address and phone number so that we can send you a reminder postcard through the mail about a week before the test that will have directions to the Sensory Evaluation Lab. If you find out you are unable to come to the test, please call us as soon as possible because it will be very important for us to schedule someone else to take your place.

7. Thank you very much for your time and participation. I hope you have a nice day.

NAME OF
INTERVIEWER_____ DATE _____

It is advisable to provide some current check during the study regarding key demographic characteristics and product usage and/or purchase-frequency information to check on current usage patterns and changes in demographic characteristics. If there is some doubt about the sample obtained through this process, alternative sampling procedures may be required (ASTM, 1979). The questions can also provide a check on the recruitment procedure for the test. Responses to the screener and additional questions may be used to update the database. Although consumer attitudes toward food are not readily altered, food consumption patterns may change due to health, nutrition, or other reasons; for this reason, periodic updating of the food-use frequency information may be warranted.

Demographic Characteristics

Demographic characteristics that may be considered and controlled, or partially controlled, in drawing a sample for consumer tests include age, gender, residence in certain geographic regions of the country, income, education and employment, and race or ethnic background. The overall objective is to select a relatively homogenous group from a fairly broad cross section, all of whom like and use the product or the product category, and exclude those individuals who exhibit extreme or unusual response patterns (Stone and Sidel, 1993).

Age and Gender

For products that have broad appeal, consumers between the ages of eighteen and sixty-four are recruited for the test. The sample should be selected by age and gender in proportion to its representation in the user population. It is important that current figures on the demographic characteristics of the target market be obtained before sampling.

Geographic Location

Regional differences in preference and usage of many products occur in various geographic areas. It is therefore necessary to test products for national distribution in more than one geographic location.

Other Consumer Characteristics

Other characteristics maybe important in describing the target market. These should be considered in the plan for recruitment of the target sample for a test.

User Group

Another primary consideration is that all the subjects should be actual or potential consumers of the product. It should not be assumed that almost any group of people is a potential consumer simply because of the broad distribution of consumption patterns that exist. Effort needs to be expended to recruit high proportions of consumers who are actual users of the product type. This can be done in various ways depending on the type of consumer test being conducted. Subjects can be screened at the time of testing as they are intercepted for a central location, mobile laboratory, or home-use test, and during the recruiting process for a laboratory test and all other tests in which prerecruitment is conducted to ensure that the product is actually used and to establish the frequency of use.

Another approach is to recognize and take advantage of the fact that product usage is often correlated with demographic characteristics. One can select a sample to recruit more of these individuals or entirely restrict recruitment to population segments that have these characteristics, such as recruiting children for a test on candy. Use of certain items is believed to be restricted to high-income households, or some products are believed to be more popular in certain regions of the country than are others. Another frequent practice in testing food products is to use panels consisting solely of the major preparer and purchaser of food in the household, even if this results in a larger proportion of female participants than in the user group, in the belief that their opinions and preferences are the main determinants of food purchases.

Employees versus Nonemployees

Often, panelists for consumer tests are recruited from within the company to participate in various tests before conducting field studies. They are laboratory, technical, or office workers who are asked to participate in the consumer panels. These individuals may have some knowledge of the product that may result in a biased response. Furthermore, they may respond in a manner that they think will please their supervisors. For these reasons, the use of employee panelists in affective tests is not recommended. A similar situation may occur in university or government laboratories where convenience samples consisting of students or staff are recruited to participate in acceptance tests. This practice is likewise not recommended for the reason that these panels may not respond in the same manner as external or "real" consumer panelists. The bias associated with the use of employees (vs. naive consumers) in a consumer sensory test should be considered in assuring validity of the sensory test. In most cases, their use should be avoided.

A trend has been to use local residents or contracting with a local agency that provides panelists for consumer acceptance tests. Although there has been an increase in the use of local residents in place of employees, it is recognized that a large number of sensory acceptance tests involve employees. While agreement exists that the use of employees in consumer affective tests is not recommended, some (Meilgaard et al., 1991) recognize that when product maintenance is the objective, company employees and those residents who reside in close proximity to company offices, technical centers, or processing plants for the product samples being tested do not represent a great risk when asked to serve on the panel. However, for new product development, product improvement, or product optimization, these individuals should not be used to represent the consumer.

The advantages of using employee panels are their availability for a longer time and obtaining a more accurate estimate of the cost of each test. On the other hand, use of employees introduces a source of bias.

Examples of Biases when Using Employees

Employees tend to favor or be more critical of products they manufacture. They either prefer products they manufacture, or if morale is bad, find reasons to reject products (Meilgaard et al., 1991). In such cases, only blind samples should be served, and products that are recognizable should be disguised. If it is not possible to disguise the products so that employees involved in their manufacture can no longer pick them out from the rest of the samples, a consumer panel should be used. Employees do not rate characteristics of products the way consumers would. Employees may have additional knowledge about their product, such as product improvement efforts by the company, that may influence their ratings. If the test product has a different appearance from the traditional product, employees may rate their acceptance of the "improved" product higher than actual consumers would. Again, this calls for presentation of samples that are disguised so that employees do not recognize them. Often, it may not be possible to disguise the product, and if so, outside testing must be used. Finally, employees do not usually represent the target segment for the product and cannot be expected to give any indication of the response of the target market. They have unusual patterns of product use because they may receive products for free or obtain these from a company store, or make selections due to brand loyalty. For these reasons, employee panels should be used with caution.

When employees are asked to participate in a sensory acceptance test, effort should be made to use the same selection criteria that would be used for a nonemployee panel. At least, employees who volunteer should be screened for product usage and their likes and dislikes.

Their food preferences should be compared with the food preferences of individuals from other populations. The data on food preferences can be obtained from published surveys but preferably from one's own surveys, and the information is used as a basis for selecting subjects for acceptance tests. Stone and Sidel (1993) propose that when employees are used, they should be likers of the product to be tested or their attitudinal score should fall ± one standard deviation of the grand mean of all survey participants.

The importance of panelist selection and frequency of test participation are critical issues in relation to the question of the objectivity and face validity of the results from employee acceptance tests. There also is a risk that employees will recognize a product despite efforts to disguise brand names or unique markings on those products.

Therefore, before making a decision on the use of results from acceptance tests involving employees, a series of comparative tests should be run to determine whether there is potential bias. These comparative test should be identical. For the analysis, Stone and Sidel (1993) and Lawless and Heymann (1997) recommend that a split-plot analysis of variance be conducted. The products are treated as the "within" variable and the panels (employee vs. nonemployee) as the "between" variable. If the F value for panels is significant, then the employee and the outside consumer panels are not similar, and the employee panel may not be substituted for the outside panel. The interaction between product and panel is significant; the panels are not rating at least one of the products similarly, and the two panels may not be substituted for each other. Both the F-value for the panels and F-value for the interaction between the panels and product must be nonsignificant for the panels to be comparable. Lawless and Heymann (1997) remind the sensory practitioner that this approach relies on acceptance of the null hypothesis to justify panel equivalence; therefore, one must be concerned about the power and sensitivity of the test. The power of the test is ensured by a large sample size. In this test, small-panel comparisons have too little sensitivity to detect any differences. Furthermore, to ensure that a difference is not missed, the p value for significance should be tested at a higher level such as 0.10 or 0.20. The Beta-risk and chances of Type II error become important in the comparison. If the employee panel is truly different from an outside panel, using the employee panel may result in serious management implications.

what's the difference?

If employee panels continue to be used, statistical validation of their results must be done periodically. Satisfactory results from these statistical analyses allow substitution of preference behavior for demographic criteria in selecting subjects for acceptance tests.

When a pool of nonemployees participate regularly in acceptance tests in much the same way as employees would, the qualifying procedures described for employees must be used, and their responses should be monitored on a continuing basis. This practice will assure maintenance of the reliability and validity of results.

Irrational Rating Behavior

Individuals who show extremes in scoring and exhibit unusual response patterns should be excluded from the list. Those individuals who rate all samples a single category or point on the 9-point hedonic scale, such as scoring products either all eights (like very much) or all twos (dislike very much) exhibit irrational rating behavior. They may do so presumably either because they do not differentiate between products, or because they are more interested in the incentive than in giving their true responses to each product, and they should be excluded from the tests. For this reason, it is a good practice to monitor responses to products and discontinue the use of these panelists.

Table 4.3 Setup and Maintenance of a Consumer Database

1. Recruitment
2. Screening and qualification
3. Collection of demographic information
4. Scheduling for a test
5. Orientation
6. Incentives
7. Monitoring panelist participation and performance
8. Panelist replacement

Consumer Database Management

A database of consumer names, demographic characteristics, product usage characteristics, and food preferences and participation in the different tests is often used as a source for potential panelists in focus group discussions and consumer laboratory tests but can be used in central location and home-use tests wherein panelists are prerecruited rather than intercepted. There are several advantages and disadvantages of maintaining a database. The consumer database is an invaluable resource for the sensory practitioner who is involved in conducting a number of consumer tests. A database provides a source for recruitment of prospective panelists with known demographic characteristics and often contains information on the food preferences and consumption frequencies of the panelist. It should likewise have information on participation of the panelist in consumer tests, and its use facilitates taking into consideration the frequency that an individual has been used in tests. The disadvantages of recruiting from a database are the need to maintain the database and the recurring need to update the database due to attrition. Selection and maintenance of a consumer panel is labor intensive. Panelists who are used frequently will become experienced test participants within a short span of time. Cooperation rate is variable and may range anywhere from 5% to 90%, and a new panel members must continually be recruited to compensate for the aging of a selected panel over the course of several years of testing, and to offset dropouts.

Setup and Maintenance

The establishment and maintenance of a consumer database include several steps from recruitment to replacement of a panelist. The different steps in the establishment and maintenance of a consumer database are listed in Table 4.3.

Recruitment

Recruitment of prospective members of a database can be from different sources. Recruitment can be conducted through intercepts at shopping areas or grocery stores, at fairs, large gatherings where consumers congregate, through newspaper or radio advertisements, from flyers placed at different locations, such as shopping areas, churches, and laundromats, from referrals from other panelists, membership lists of various organizations, calling consumers at random from a telephone directory, or by using random-digit dialing.

Screening and Qualification

Initially, only a few criteria are used to screen prospective members of a consumer database. Interest of the consumer is the primary factor. In other cases, age may be a consideration, and recruiters may limit the age of the panelists to eighteen to sixty-four. A number of databases only include one member of a household, usually the major preparer and purchaser of food. If only family members are needed in a test, the numbers and ages of different household members are included in the database; they can be recruited through the participant. Often, recruiters ask the consumers whether they are the major preparer and purchaser of food in the household.

Collection of Demographic Information

The collection of demographic information requires the design of a questionnaire such as that in Figure 4.1. The questionnaire should contain all the information that will be necessary for prescreening a prospective participant. If this information is obtained and updated, it becomes a time-saving resource for recruiters. The data collection can be by face-to-face interviews or by self-administered questionnaire. A minimum amount of data is collected at the recruitment site: names, address, telephone number and only a few demographic questions such as age, gender, race, and employment status. Additional data will be collected during the follow-up interview by telephone or through a self-administered questionnaire.

The data are entered using a generic database management software program such as Filemaker Pro or other specialized consumer panelist database management software such as PanMan. Computerization of the database helps to maximize the efficiency of using the database during recruitment and also in database maintenance.

Scheduling Consumers for a Test

When it is necessary to recruit for a consumer test, the recruiter uses the database to prescreen for prospective panelists using database information, then

Panelist Code # _____

For Office use only

Panelist Code #

1 2 3 4 5 6 7 8 9 0

Please answer all the questions. Your name is not on the questionnaire and will not be identified with your answers.

Q-1 We would like to begin by asking if you are the person who buys and cooks the food in your household most of the time?

Yes ☐   No ☐

Q-2 In your household, the person who buys food and cooks most of the time is the:

Husband ☐   Wife ☐   Child ☐   Self ☐   Other ☐

Q-3 We want to ask you a few questions about yourself. This information will help us compare opinions of oeople with different backgrounds.

What is your date of birth? _____; What is your age group?

18-25 ☐   25-34 ☐   35-44 ☐   45-54 ☐   55-64 ☐   >64 ☐

What is your sex?

Male ☐   Female ☐

Which do you consider yourself to be? (Check one).

White ☐   Black ☐   Spanish/ Hispanic ☐   Other, Specify _____ ☐

What is your marital status? (Check one)

Never Married ☐   Married ☐   Separated ☐   Divorced ☐   Widowed ☐

Please mark how many people in each age group live in your household. Include yourself in the count.

	0	1	2	3	4	5	6	7	8	>8
Under 6 yrs. old	☐	☐	☐	☐	☐	☐	☐	☐	☐	☐
7-12 yrs.	☐	☐	☐	☐	☐	☐	☐	☐	☐	☐
13-18 yrs.	☐	☐	☐	☐	☐	☐	☐	☐	☐	☐
19-24 yrs.	☐	☐	☐	☐	☐	☐	☐	☐	☐	☐
25-64 yrs.	☐	☐	☐	☐	☐	☐	☐	☐	☐	☐
Over 64 yrs.	☐	☐	☐	☐	☐	☐	☐	☐	☐	☐
Total Number of people in household	☐	☐	☐	☐	☐	☐	☐	☐	☐	☐

If your household has any children under 19, how many parents are living in it?

One parent ☐   Both parents ☐   No parents ☐

continued

Figure 4.1   An example of a demographic questionnaire for consumer database participants.

For Office use only

Panelist Code #

How far did you go in school?
- [ ] less than seven years of school
- [ ] junior high school
- [ ] some high school
- [ ] completed high school or equivalent
- [ ] some college
- [ ] Graduate or professional school

Please check the one which best applies to you:
- [ ] Employed full time
- [ ] Employed part time
- [ ] Home maker
- [ ] Student
- [ ] Retired
- [ ] Unemployed
- [ ] Disabled

How many people in your household contribute to the household income?

	0	1	2	3	4
Number of people employed full time (check one)	[ ]	[ ]	[ ]	[ ]	[ ]
Number of people employed part-time (check one)	[ ]	[ ]	[ ]	[ ]	[ ]
Number of people with other sources of income (pensions, social security etc). check one:	[ ]	[ ]	[ ]	[ ]	[ ]

What was the approximate income level of your household last year? (Check one).

under $9,999	$10,000-$19,000	$20,000-$29,999	$30,000-$39,999	$40,000-$49,999	$50,000-$59,999	$60,000-$69,999	$70,000- and over
[ ]	[ ]	[ ]	[ ]	[ ]	[ ]	[ ]	[ ]

Check the item that best describes the job of the head of your household. (If retired or not now working, describe the former or usual job). If you are the head of the household, check the item that describes your job.

- [ ] Executive or proprietor of large concern
- [ ] Manager or proprietor of medium concern
- [ ] Administrative personnel of large concern or owner of small independent business
- [ ] Owner of little business establishment, clerical, sales work or technician
- [ ] Skilled worker
- [ ] Semi-skilled worker
- [ ] Unskilled worker

If not sure, describe the job:
Title _____
Kind of work ___

continued

Figure 4.1 *Continued.*

For Office use only

Panelist Code #

Do you own or rent the place where you live?
- ☐ Own (or buying)
- ☐ Rent
- ☐ Other

In what type of home do you live?
- ☐ One family house
- ☐ Mobile home
- ☐ Apartment, condominium, duplex, or townhouse

How many years have you lived in Georgia? _____ years

For office use only
years

In Georgia, people live in a city and a county. Please give the name of your city _____

For office use only
City name

Please give the name of your county _____.

For office use only
County name

Q-4 We want to ask you some questions about your shopping and eating habits. How many times per month do you shop for groceries? _____

For office use only
times per month

continued

Figure 4.1 *Continued.*

For Office use only

Panelist Code #

= = = = = = = = = =
= = = = = = = = = =
= = = = = = = = = =
= = = = = = = = = =

1 2 3 4 5 6 7 8 9 0

How many different grocery stores do you regularly shop at (at least once per month)? _____

For office use only
stores per month

= = = = = = = = = =

1 2 3 4 5 6 7 8 9 0

How many times a day do you generally have something to eat (include snacks and meals)? _____

For office use only
eat

= = = = = = = = = =

1 2 3 4 5 6 7 8 9 0

On how many of these occasions during the day do you eat alone? _____

For office use only
eat alone

= = = = = = = = = =

1 2 3 4 5 6 7 8 9 0

Approximately how many times do you eat out in one week? _____

For office use only
eat out

= = = = = = = = = =
= = = = = = = = = =

1 2 3 4 5 6 7 8 9 0

How many of these are take-out meals? _____

For office use only
take-out

= = = = = = = = = =

1 2 3 4 5 6 7 8 9 0

How tall are you? _____ feet _____ inches

For office use only

ft

= = = = = = = = = =

1 2 3 4 5 6 7 8 9

inches

= = = = = = = = = =
= = = = = = = = = =

1 2 3 4 5 6 7 8 9 0

continued

Figure 4.1 *Continued.*

For Office use only

Panelist Code #

What is your present weight? _____ lbs

For office use only
take-out

Check all that apply to your present weight
- [ ] I want to lose weight
- [ ] I do not want to gain or lose weight
- [ ] I want to gain weight

How many pounds do you want to lose or gain? _____ lbs

For office use only
take-out

Check the statement that comes closest to describing you at the present time.
(Check one only)
- [ ] I am on a diet right now.
- [ ] I diet from time to time not now.
- [ ] I don't diet but I never eat certain fattening foods.
- [ ] I cut down somewhat on fattening foods.
- [ ] I sometimes cut down on fattening foods for a few days but in general pay no attention to my weight.
- [ ] I eat all I want and am not concerned with putting on weight.

Figure 4.1 *Continued.*

uses a screening questionnaire that has been prepared for the specific test to be conducted. It makes good sense at this point to recruit all new consumers in the database to a test. If consumers do not qualify, they should be scheduled for the next consumer test for which they qualify. Participation in a consumer test soon after recruitment helps to maintain their interest in participating in consumer tests.

Orientation

Consumers, especially those participating for the first time in a consumer test, need to be oriented. The orientation may be on the use of signal lights in the sensory booth area or the test procedure. The orientation should be as short as possible to accomplish the objective, and should be given prior to the test. Preferably, recruitment for a test should screen out consumers who have participated in a consumer test in the last six months. If this practice is upheld, the orientation should be given to all participants, not just those who are participating for the first time.

Incentives

Incentives help to maintain the interest of consumers in participating in the tests. The incentives may vary according to the time spent during the test and the number of tasks required.

Monitoring Panelist Participation and Performance

One of the most important requirements in the maintenance of a database is monitoring participation and performance. Participation is important in the test to obtain the number of needed responses. The percentage of panelists scheduled for a test that show up for a test should be tracked. This number should be used in recruiting panelists for the test. For example, when the "no-show" rate is 20% in a test requiring 100 responses, it would be prudent to recruit 120 panelists for the test. Recruiting takes much time, and an increase in the "no-show" rates increases this further. When consumers recruited from a database do not have a good reason for not showing up for a test, they should be placed in the inactive list of participants. The inactive list includes those who have moved or asked to be removed due to change in employment status or other reasons. It is a waste of time, labor, and resources, and it does not make good sense to recruit prospective panelists who do not have the sense of commitment to show up when scheduled for a test.

Performance is likewise important. The sample servers and project leader should be observant in monitoring panelist performance. Panelists who appear to have comprehension problems when no one else in the panel has these, those who do not follow instructions, those who are anxious about the testing method and push the call button for help more often then others, as well as those who exhibit unusual scoring patterns (scoring one point on the hedonic scale for all samples and all attributes in the set) should be placed in the inactive list. Reasons for being on the inactive list should be noted on the database.

The amount of information and updates on a consumer in the database is important to conducting a valid and reliable test. Participation and performance of the panelist are important to track and should be clearly marked in the database if these are not satisfactory. For this reason, the use of a central database during recruitment enables the recruiter to assemble the most appropriate panel for a given test.

Panelist Replacement

Panelists on the database will need to be replaced at some time. Reasons for replacement include age, illness, death, loss of interest, change in employment status, and change of address to outside the testing area. These panelists are placed on the inactive list with the reasons for replacement.

The entire database of consumer characteristics should be filed in a computer. This database is used to select prospective panelists for a test, taking into consideration the frequency. As tests are done, updates of the consumer files are entered.

Nonusers of the product should be used only when there is a compelling reason why a nonuser should be used, such as when testing an entirely new product and there is a no user group established for the product.

Special Problems in Sampling

Some special problems are encountered in obtaining samplings of subjects for central locations tests. Theoretically, it might be possible to achieve as representative a sample as in any other survey; however, the desire for speed and lower costs usually requires a greater degree of compromise. The location or locations that are used usually constitute one major factor, because only certain types of people will be available.

In drawing a sample one must consider that the sample should form a fairly broad cross section, not weighted heavily by a single population segment, such as males versus females, elderly versus younger participants, and low versus average socioeconomic status, and so forth. This is often no more than a matter of protection, since it has been found that, in general, preference direction is approximately the same for all segments of American society—therefore, even though it is seldom possible or practical to obtain a representative cross section of subjects, experience has shown that it is satisfactory to use a reasonable good sampling of available consumers.

Use of Trained Judges in Affective Tests

Individuals who are qualified for discrimination and descriptive tests should not be used as participants in consumer acceptance tests, regardless of their willingness to participate. Panelists who are used in sensory discrimination or descriptive tests have been selected for their skill in performing sensory evaluation tasks or have become skilled as a result of undergoing training. The training process results in individuals who have an analytical approach to sensory evaluation that will bias the overall responses required for the acceptance task (Stone and Sidel, 1993). Once trained, these individuals no longer represent the target market or potential users of a product. Similarly, those individuals who possess technical or related information about a specific products should not be used because of their potential bias (Stone and Sidel, 1993). Objectivity of the panelists needs to be maintained if results are to be considered reliable and valid (Stone and Sidel, 1993).

Discrimination Testing to Screen Consumer Panelists

Individuals who are qualified for discrimination and descriptive tests should not be used for acceptance tests, regardless of their willingness to participate. The description of the discrimination test used by Morrison (1981), wherein he cites Moskowitz (1980), pointed to the continued interest in use of the discrimination model to screen prospective panelists. Stone and Sidel (1993) pointed out a basic flaw in this approach, stating that subjects would require a considerable number of trials, usually more than ten, before a discriminator or nondiscriminator can be identified. It is therefore an inappropriate procedure for measuring preference. Because the discrimination task is analytical and the preference test is affective, naive consumers cannot be expected to handle both without considerable confusion and "most likely a confounding of the preference judgment." (Stone & Sidel, 1993)

Garbage in, Garbage Out

"No amount of sophisticated statistical analysis will make good data out of bad data" (O'Mahony, 1986). All the elaborate statistical techniques will not mask the deficiencies of an inappropriately drawn sample.

References

ASTM, Committee E-18. 1979. *ASTM Manual on Consumer Sensory Evaluation*, ASTM Special Technical Publication 682, E. E. Schaefer, ed. American Society for Testing and Materials, Philadelphia, PA, pp. 28–30.

Chambers, E., IV, and Smith, E. A. 1991. The use of qualitative research in product research and development. In *Sensory Science Theory and Applications in Foods*. H. T. Lawless, and B. P. Klein, Eds. Marcel Dekker, New York, Basel, and Hong Kong. pp. 395–412.

IFT/SED. 1981. Sensory evaluation guide for testing food and beverage products. *Food Technol.* 35(11):550–559.

Lawless, H. T., and Heymann, H. 1997. *Sensory Evaluation of Food: Principles and Practices*. Chapman and Hall, New York.

Meilgaard, M., Civille, G. V., and Carr, B. T. 1991. *Sensory Evaluation Techniques*, 2nd ed. CRC Press, Boca Raton, FL.

Morrison, M. G. 1981. Triangle taste test: Are the subjects who respond correctly lucky or good? *J. Market.* 45:111–119.

Moskowitz, H. R. 1980. Product optimization as a tool in product planning. *Drug Cosmet. Ind.* 126(6):48, 50, 52, 54, 124–126.

O'Mahony, M. 1986. *Sensory Evaluation of Food*. Marcel Dekker, New York, pp. 8.

O'Mahony, M., Thieme, U., and Goldstein, L. R. 1988. The warm-up effect as a means of increasing the discriminability of sensory difference tests. *J. Food Sci.* 53:1848–1850.

Sidel, J. L., and Stone, H. 1976. Experimental design and analysis of sensory tests. *Food Technol.* 30(11):32–38.

Sokolow, H. 1988. Qualitative methods for language development. In *Applied Sensory Analysis of Foods*, Vol. I., H. Moskowitz, ed. CRC Press, Boca Raton, FL. pp. 3–19.

Stone, H., and Sidel, J. L. 1993. *Sensory Evaluation Practices*, 2nd ed. Academic Press, San Diego, CA.

5

Qualitative Methods—
Focus Groups

Introduction

Consumer sensory research can be classified into two major categories: qualitative and quantitative. Quantitative research involves measurements, whereas qualitative research is descriptive and does not involve measurements. Qualitative consumer research methods are useful in defining critical attributes of a product. Several qualitative methods exist, and these include one-on-one, indepth interviews, group interviews, and focus groups. The most commonly used qualitative research method is the focus group.

The focus group is a method by which small groups of consumers are used to obtain information about their reaction to products and concepts, and to investigate various other aspects of respondents' perceptions and reactions. This method is used to determine product attributes that consumers think are important and should be maximized in the product, and characteristics that consumers do not like and think should be minimized or eliminated from the product. The distinguishing feature of this method is the unstructured approach.

Reasons for Conducting Focus Groups

Optimization of a Product's Acceptance

Product optimization requires the identification of the sensory properties important to consumer acceptance (Schutz, 1983). One means to determine the critical attributes of a product is through qualitative research. Qualitative consumer research methods can be used to investigate a wide range of issues and obtain detailed information about consumer attitudes, opinions, perceptions, behaviors, habits, and practices (Chambers and Smith, 1991). One of the most frequently used qualitative methods is focus group research (Greenbaum, 1988). It is useful in gaining insight into consumer's preferences and defining critical attributes of a product (Galvez and Resurreccion, 1992). They may also be used in studying consumer habits (Hashim et al., 1995) or attitudes (Hashim et al., 1996), which may be predictive of future behavior.

Early Assessment of a Concept or Prototype

The focus group method is often used in the very early assessment of either a concept or a prototype (ASTM, 1979). It utilizes small groups of 8–12 consumers to obtain information about reactions, both positive and negative, to products or concepts, and to investigate various other aspects of respondents' perceptions and reactions. It is useful in determining ways a product can be used; words, feelings and emotions that arise when consumers talk about the products; product attributes that consumers think are important and should be maximized in the product; and characteristics that consumers do not like and think should be minimized or eliminated from the product; consumer perceptions about the sensory properties of the product; overall, variations in taste and flavor, texture, and so on. Product prototypes or mock-ups may be introduced, or there may be evaluation of a finished product or a competing product. The primary interest in conducting a focus group is in generating the widest possible range of ideas and reactions—more than attempting to get definite information on any specific points.

Facilitation of Quantitative Research

Although quantitative research may provide answers that are projectable to a larger population, it may not be able to give the type of responses that focus groups can provide. Qualitative research usually needs to precede quantitative testing to help establish criteria for data collection and to follow quantitative testing to aid the explanation of quantitative data (Chambers and Smith, 1991). Focus groups conducted before quantitative research will help define the problem. If this knowledge does not exist, a great deal of time and money can be wasted on research that yields no actionable data.

Gathering Extensive Information about a Product Category

Focus groups may also be useful in gathering extensive information about the product category. For example, a producer of soup may be interested in extending its product line to include frozen soups for lunches. It only has a sketchy idea on how this product category will be perceived, so it conducts focus groups to help make decisions on how to position the product. The focus group can also help to generate information on the language that consumers use to talk about a product category.

Company employees often discuss products using industry jargon, which

Table 5.1   Advantages of the Focus Group Technique

1. Flexibility.
2. Provides observation of real consumers in an interactive setting.
3. Involves fewer participants compared to quantitative methods.
4. Can be arranged on short notice and at a lower cost.
5. Statistical analysis is unnecessary.

what's one of those?

may confuse or mislead consumers. Their use in a questionnaire may likely elicit misleading data and erroneous interpretation. For example, the use of the terms _radiated_ chicken versus *irradiated* chicken would result in a negative consumer reaction to chicken treated with ionizing radiation for microbiological safety. Finally, focus groups can be useful in designing a more valid questionnaire by helping to determine the most important questions and appropriate wording on the questionnaire. However, it must be remembered that while focus groups may offer, in some instances, faster turn-around time and a lower cost one than large-scale quantitative surveys, and provide results that prove to be useful, the focus group should not be used as a substitute for quantitative research.

Advantages and Disadvantages of Focus Groups

Advantages of Focus Groups

There are several advantages for using focus groups to obtain consumer information not easily obtained with quantitative research. The advantages of conducting focus groups are listed in Table 5.1.

Focus Groups Offer Flexibility

Because of their flexibility, focus groups are able to meet many diverse marketing and research and development (R&D) objectives including idea generation, testing new products and new product concepts, commercials, and questionnaire development for quantitative research. Furthermore, the flexibility of the method allows for new objectives and insights as the discussion progresses. Focus groups provide the ability for marketing, R&D, and management personnel to observe real consumers in an interactive setting. In a focus group, observers can easily view interactions between the consumer and the new product or concept. Furthermore, observers can view participants interacting with other participants as they build upon each other's observations and input. Focus

Table 5.2 Disadvantages of the Focus Group

1. Focus group results are not quantitative.
2. Sample size is small; results are not projectable.
3. Participants do not represent the target market.
4. Topics and direction of the discussion are moderator dependent.
5. Careful interpretation of the data is crucial.

groups involve fewer participants compared to quantitative methods. This results in considerably less effort in the recruitment of qualified consumers who represent the target market for the product or product concept. Participants do not need training; thus, focus groups can be arranged on short notice and at a lower cost relative to a large-scale study. Because the method is qualitative, statistical analysis is inappropriate and unnecessary.

Disadvantages of the Focus Group

Although the focus group presents numerous advantages, there are several disadvantages. The disadvantages of the focus group are listed in Table 5.2.

Focus group results are not quantitative. No measurements are made, and statistical treatment of qualitative data is inappropriate. Results cannot be used independently of quantitative research for strategic decision support. The sample size is small, consisting of 8–12 participants per session; therefore, results are not projectable to a population. Although participants are screened for their usage patterns for the product, and demographic characteristics, participants do not represent the target market. Although a serious attempt should be made to screen for participants that represent the target market, the small sample size does not allow the target market to be represented. The topics and direction of the discussion are moderator dependent. Careful interpretation of the data is required; management should understand the scope and relevance of the method.

Role of the Project Leader

The project leader is responsible for the entire test, from the planning stage to the reporting of final results to the client or requester. The project leader coordinates and organizes the study. The project leader needs to thoroughly understand the purpose and goals of the study as stated by the client. With this understanding, the project leader must develop a focus group plan for the client that can deliver the information needed by the client, and outlines assumptions expected from the client and information that can be delivered as a result of the study.

The project leader has several responsibilities during the planning and conduct of a focus group. The major areas of responsibility for the project leader are as follows:

1. Responsible for planning and implementation of the study.
2. Describes the focus group participants, panel size, and classification quotas for recruitment.
3. Lists screening criteria; prepares recruitment script; screens and monitors recruitment performance.
4. Prepares Moderator's Guide after brainstorming with requestor of the test and key project personnel to ensure that all issues of interest are covered.
5. States focus group site requirements, including special equipment needs, and provides suggestions for participant incentives.
6. Greets panelists.
7. Observes the focus group if the moderator is a different individual
8. Transcribes tapes.
9. Listens to the tapes and interprets results.
10. Conducts debriefing.
11. Writes report.

The leader is responsible for planning and implementation of all components of the study such as recruitment, including determining criteria for participant selection, development of the recruitment screener, and obtaining demographic information to minimize the length of time required for the conduct of the focus group. The project leader develops a guide after brainstorming with the requester of the test and other project personnel. The discussion is conducted to ensure that all issues that are important to the accomplishment of the project objectives are discussed. The project leader then drafts the guide and submits it for revision. The project leader makes logistical arrangements and is responsible for the focus group facilities, including arrangements with hotels for use of conference rooms when their use is necessary. The project leader collects and prepares samples, and completes all paperwork according to schedule. The project leader makes certain that all equipment, such as audio- or videotaping equipment, slide or overhead projectors, and other audiovisual equipment such as video players and video monitors, are in good working order, set up, and ready to use during the test. On the day of the focus group, the project leader makes sure that panelists are greeted and that their incentives are ready for disbursement. If the project leader is a different individual from the moderator, the project leader observes the focus group. After the study, the project leader has the tapes transcribed and should listen to the tapes. The project leader interprets results and conducts the debriefing with the client, then writes the report.

Role of the Moderator

The focus group discussion group is conducted under the guidance (ASTM, 1979) of a qualified focus group moderator (Wu, 1989). The panel moderator occupies a key role in the focus groups. The role of the moderator is to

1. Introduce topics and leads the discussion.
2. Create an atmosphere conducive to obtaining panelists' reactions.
3. Keep discussion focused on issues of interest.

The functions of the panel moderator are to introduce topics and lead the discussion; recognize important points and encourage the group to explore and elaborate on them; observe the nonverbal communication among respondents, between respondents and moderator, and between respondents and subject matter; create an atmosphere that allows respondents to relax and lower some of their defenses; and synthesize the understanding gained by the objectives. The moderator must make sure that all participants express their opinions, prevent participants from monopolizing the discussion, assure that the discussion time is used efficiently, and make sure that the discussion centers on the main issues of interest.

The moderator may vary from a technician trained on the job to a psychologist skilled in group dynamics. It is not necessary for the moderator to have specialized educational training in psychology or sociology in order to achieve value from a group session. However, suitable training on focus group moderation will provide information on development of a Moderator's Guide, probing techniques, meeting research objectives, leading discussions, interpreting results, and writing reports. Good moderating skills are developed with practice and training.

Desirable Characteristics of a Panel Moderator

Desirable characteristics of a panel moderator are as follows:

1. Qualified and objective
2. Friendly and a good listener
3. Can handle diverse opinions and personalities
4. Flexible and responsive
5. Assertive but diplomatic

6. Maintains a good flow of information
7. Recognizes and guards against moderator bias

A panel moderator should be qualified and objective, with no vested interest in the outcome of the study. Furthermore, the moderator should be friendly, empathic, sensitive, a good listener rather than informer, and nonjudgmental. A good panel moderator must have the skills to be able to handle diverse opinions and personalities without becoming judgmental and be able to establish a nonthreatening atmosphere so that participants should be able to state their perceptions, opinions, beliefs, and attitudes without fear of criticism. An effective moderator will be flexible and responsive to quickly modify the questioning to explore ideas generated early by the group that the client may be interested in. The moderator should be assertive but diplomatic, able to tactfully place reasonable limits on each person's participation, and control the discussion so that multiple panelists do not speak at the same time. The moderator should try to maintain a good flow of information and to limit intersubject discussion when there seems to be danger of destroying continuity. The moderator should not appear to be an expert on the topic of the discussion or discussion may be reduced. Many users of focus group information may prefer to have a moderator who is technically uninformed about the issues and has no preformed opinion. While this helps to ensure an unbiased discussion and report, it can result in the loss of an opportunity to probe important technical issues that may arise (Lawless and Heymann, 1997).

Moderator bias may easily occur. Undue support may be given to ideas of participants one agrees with by giving more eye contact, verbal affirmation, head nodding, calling on them first, or being more patient with them (Kennedy, 1976). Opinions that are unwanted can be deemphasized by lack of follow-up, and failing to probe, summarize, or reiterate contrasting or unfavorable opinions (Lawless and Heymann, 1997). A conscientious moderator will recognize these behaviors and evade them.

Panel Moderation

The moderator must clearly understand the objectives of the project and know enough about the project to facilitate general and detailed discussion. The project leader, if other than the moderator, moderator, and the client should brainstorm and then review the Moderator's Guide together, preferably in a "dry-run" prior to the focus group to ensure that the moderator demonstrates understanding of the objectives of the focus group and the client understands the flow of the discussion topics. The guide should be checked to ensure that objectives will be met, relevant topics are covered, and key issues are addressed.

All products, tests and alternates or substitutes, or concept boards should be used during the "dry-run." The moderator must be responsible for determining the actual flow of the discussion to ensure that the progression seems natural. Rapport has to be established prior to in-depth questioning that takes place. The moderator must allow for diverse opinions and accept lack of consensus or closure. When discussion goes off-tangent, the moderator firmly redirects without losing group affinity, and must be able to do linking and logical tracking to be sure all key points have been explored and summarized as well as possible.

Observers

Client observers should not disrupt the group process. Although an occasional note from the client is permissible, communication between the moderator and the client should be infrequent and not suggest to participants that an outside party is directing the discussion. Clients usually observe the focus groups in an observation room next to the focus group room, separated only by an interior wall with a one-way mirror; it is important to remind panelists not to make noises audible to respondents. Furthermore, conversations by clients should be limited, and phone calls should not be permitted until the focus group has been adjourned. Clients may make notes on their copy of the Moderator's Guide. Observers should focus on what the consumer is trying to communicate, verbally and nonverbally, through body language and facial expressions, and not be distracted by speaking skills. They should be objective and receptive to new ideas rather than attending for the purpose of confirming their opinions.

Participants

Focus groups vary in size and ideally consist of 8–12 participants who represent the projected target market and are interviewed in depth (ASTM, 1979; Sokolow, 1988; Chambers and Smith, 1991). It is, however, difficult to draw a focus group sample that is truly representative of typical consumers or anticipated consumers of the product type involved. The selection of panel members is sometimes more rigorous, but in most cases those who can be most conveniently contacted are utilized. Fewer than 8 participants do not provide adequate input, whereas greater than 10 participants do not provide sufficient time during a one-and-a-half-hour to two-hour session to cover the number of issues desired.

Consumer Sample

In focus groups, one should use a judgment sampling as much as possible. The project coordinator should make sure that those invited to participate in the group meet the specifications. Frequently, regional differences are thought to exist, but because of the mobile population they usually do not. The assumption is usually made that a small group can generate ideas that can then be checked out further in quantitative research. The need for a judgment sample arises from the desire to have ideas come from the proper group; for example, if the target population is that of elderly consumers, there should be no one in the panel that is less than fifty-five years of age.

Recruitment

Recruitment of respondents may be done through market research agencies or by the sensory group, through telephone calls or personal contacts. Methods often used are random selection from a telephone directory, random-digit dialing, posters in retail stores, referrals, mailing lists from organizations, consumer databases and intercepts at malls, shopping areas, or restaurants. When the research facility does not maintain a consumer database, it is common practice to seek the help of various social organizations to provide lists of members or do the recruiting. Some examples of consumer groups contacted for the studies are churches, schools, or professional groups, and women and men's clubs. Attempts should be made to recruit participants who are representative of typical consumers of the product type or be anticipated users of the product. When the possibility of regional differences in reactions is suspected, sessions may be conducted in different parts of the country.

Screening

Participants usually need to qualify according to predetermined criteria that describe the target market for the product. A screener should be developed collaboratively with the sponsor of the test. The screener should be prepared so that it ensures reliability of the screening process and provides a tally for problem of having too few responses to verify quotas. Selection of participants should be done in a rigorous way, although, at times, the undesirable practice of using those participants who can be most conveniently contacted, such as those people who live close to the facility, is utilized. Demographic criteria such as age, sex, frequency of product use, availability during the test date, allergies, and other criteria such as employment with the sponsoring company or other security screening criteria are often used. In addition, ethnic or cultural background, occupation, education, family income, and experience and/or participation in focus groups, surveys or sensory tests may be used.

It is advisable to avoid recruiting "professional" focus group participants, consumers who have poor verbal skills, and friends or relatives of project staff. These individuals would bias the results of the discussion and should not be used. "Professional" participants are those who do not fit the criteria in the screener but respond to the screener to qualify as a participant. These individuals have participated in several tests and are no longer naive consumers. Respondents with poor verbal skills should be eliminated during screening. Poor verbal skills will lead to lack of input during the focus group discussions. Friends or relatives of project staff may tend to present only socially acceptable comments.

Regional Differences

When the possibility of regional difference exists, sessions may be conducted in several different parts of the country. It is advisable to hold at least three focus group sessions to determine reliability of responses among groups.

Attendance

The problem of "no-shows" should be avoided. It reduces the validity of the focus group and is a major cause for embarrassment with client observers. To minimize the attendance problem, a number of steps can be taken. Select participants who live no more than thirty minutes from a conveniently located focus group facility, and give participants clear directions and a map to help them find the facility; send reminder letters or postcards, pay adequate fees, overbook; make participants understand the importance of their attendance and promptness.

Incentives

Participants in focus groups are usually paid for their time and effort. The amount of the incentive paid to participants may differ according to many factors, including location of the test, incidence rates of qualified participants, travel costs of participants, and length of the session.

In general, payment is given after the session is completed. Incentives for participation are provided in the form of cash, selection from a gift catalogue, gift certificates, and tickets to special functions such as ball games or concerts, or donation to charity or a nonprofit organization.

Physical Facilities

Location of Facility

The focus group needs to be held in an area of town where participants will feel safe. The facility should be easy to find and convenient for participants. The building should be easy to find, with a clearly marked entrance. Sufficient parking should be available. If the focus groups are scheduled in the evening, as most focus group sessions are, the parking area must be well lit.

Focus Group Facility

Discussions should be isolated from the interference of outside activities and events. The room should preferably not be on the street side of the building for two reasons: Noise from passing traffic could distract participants and also be picked up on tape and interfere with transcription. In addition, the room should be free from distractions such as telephones and office equipment, noisy lobbies, or hallways. Food preparation facilities should not be visible from the room. The room should be large enough to seat participants comfortably but not be so large that participants feel they are in a hall. Lighting should be adequate for reading, writing, and examination of products or concept boards. The temperature should be comfortable and controlled to maintain temperature constant for 8–10 participants for the duration of the discussions. The room must be clean and free from odors. Facilities may include an observation room with a one-way mirror, and audio monitoring and soundproofing, so that observers do not interfere with the proceedings and inhibit or distract participants. Adequate writing space for note taking is recommended.

Furnishings

Furniture must be comfortable and functional. Usually, focus group participants sit around a table so that they can have a surface to write on or to examine products. It is necessary to have a table large enough to seat all participants so that moderator can see everyone with ease during the discussions and also be able to read the name of each participant seated around the table; otherwise the shape of the table is a matter of personal choice and availability.

Reception Room

Ideally, participants should be able to go directly to the focus group room, hang up their coats, and use the remaining time to become comfortable. If the focus group room is unavailable, or if the participants have to eat before the discussions, the participants should be received in a separate room large enough for twelve participants to sit, read magazines or other material, or eat. If replicate focus groups are conducted on the same day but scheduled one after the other, focus group participants should not be held in a room where they can hear comments from other participants from other focus group sessions.

Other Considerations

Food

When focus groups are scheduled around mealtime, it is often the custom to provide food to participants. If food is served, this should be done after the discussion group has been adjourned and audio equipment has been put away. If food is served prior to the focus group, care should be taken to ensure that the selection of food or beverage does not interfere with the test. Alcoholic beverages should not be served as refreshment.

Structure

The emphasis is on ease and informality. The method is designed to capitalize on the concept of group dynamics and provides an in-depth understanding of attitudes, perceptions, opinions, and beliefs. The interaction between participants often generates more discussion and ideas than when an individual participant reacts to questions.

Length of Sessions

The sessions usually last for one and one-half hours. Few sessions are less than one hour; many last almost two hours.

Taping of Sessions

All sessions are audiotaped to provide a record of the deliberations that will be used in the interpretation of focus group results and preparation of the report. Almost all focus group sessions are audiotaped. Audiotaping does not induce self-consciousness, nor does it appear to inhibit the flow of conversation. More recently, videotape recording has been increasingly utilized. The record of facial expressions, gestures, and body movements assists in the verbal record (ASTM, 1979). It also enables identification of the source of comments. Use of the videotapes provides an alternative to using observers whose presence may inhibit free discussion or be distracting to consumers.

The participants should be informed during the initial part of the session that the session is being taped and in most cases will have signed a consent form for their participation and to be audio- or videotaped. The audiovisual equipment should allow for clear recording. Often, two tape-recorders are used to provide back up.

Props

If the discussion is on package design, actual product use, or evaluation, carefully organize these and code with three-digit codes, if needed, before the study.

If the item for discussion is visual, have several copies or items to pass around. If the discussion is on food products, have all supplies such as water, napkins, plates, eating and serving utensils ready. When discussing an item, be sure the moderator refers to it often by number to aid in transcription of the tapes. The test products, prototypes, or mock-ups, and, if available, an array of similar products are presented to the panel, allowing them to see and touch the product. "Situations such as these may require the input of a sensory evaluation expert to guard against any possible biases and to help interpret the often obtuse comments by naive subjects describing their perceptions" (ASTM, 1979).

Focus Group Procedure

Although focus group are loosely structured, the format takes the structure described: In a focus group, 8 to 12 people sit around a table, with a moderator who leads the discussion. The discussion lasts from 90 to 120 minutes.

The distinguishing feature of this method is the unstructured approach (ASTM, 1979). Flexibility and permissiveness are key elements; hence, interaction among subjects and innovative ways of responding are prized (ASTM, 1979). By careful probing of opinions about the need, critical attributes, and potential marketability of the concept or product, the concept or product can be qualitatively assessed. Labels (Teague and Anderson, 1995) or advertising may also be evaluated. In all cases, primary interest lies in generating the widest possible range of ideas and reactions. This is considered more important than attempting to get definite information on any specific points.

When a concept is determined to be worth further development, focus group sessions are a useful means to guide further development. The test products and similar or alternate products, if any, are presented to the consumers for examination.

The focus group takes place in several phases, as illustrated in Figure 5.1. These are the introduction, rapport/reconnaissance, in-depth investigation, and closure (Galvez and Resurreccion, 1992).

Introduction

The group begins with the introduction phase that is design to last approximately ten minutes. Ordinarily, this is a type of warm-up in which the general-purpose nature and the purpose of the discussion are described. The role of the moderator is explained to the participants at this time. The objectives of the specific focus group are explained to the participants. Ground rules for the discussions are likewise discussed. Such ground rules would include the following: informing the consumers that the discussion is being audiotaped or videotaped, asking

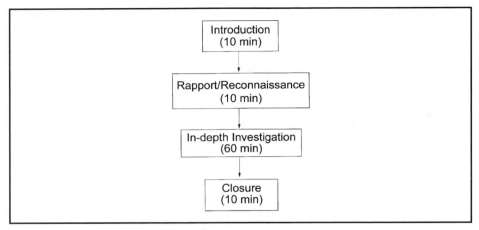

Figure 5.1   Focus group flow diagram (Galvez and Resurreccion, 1992; Reprinted with permission).

the participants to speak loudly so that their responses can be recorded and asking participants to state their name before speaking.

Participants are cautioned about the fact that there are no right or wrong answers and their honest opinions are desired. This is done to minimize pleasing the moderator and to avoid the tendency to parrot other panelists. During this phase, the moderator asks everyone to introduce him- or herself. This may be done by taking turns around the table, with the moderator introducing himself first. The introduction may take place in ten minutes.

Rapport/Reconnaissance

In the rapport/reconnaissance phase, the discussion moves to general issues. Examples of topics covered during this phase are general food consumption habits and practices of participants. Often, the critical issues will arise at this stage. The moderator uses a guide (Moderator's Guide) that directs the flow of the discussion and ensures that all the important issues are discussed. An example of a Moderator's Guide is outlined in Table 5.3. This phase may take approximately ten minutes.

In-Depth Investigation

An in-depth investigation follows, lasting approximately sixty minutes.

Closure

At the end of the discussion, the moderator closes the discussion. Closure may take place in ten minutes.

Table 5.3 Moderator's Guide: Focus Group on Packaging of Fluid Milk

I. INTRODUCTION (10 minutes)
  A. Moderator's introduction
    1. General Nature and Purpose of a Focus Group
      Introduction
      Definition and purpose of focus group
    2. Role of the Moderator
  B. Objectives of this focus group
  C. Ground rules
    Mention taping, volume of voice when speaking
    State name each time one speaks, confidentiality
    No interruptions when someone is speaking
    Avoid side conversations with other participants
    Mention incentive
  D. Self-introductions
    Name
    Area of residence
    Occupation
    Family information
    Hobbies
II. RAPPORT/RECONNAISSANCE: (20 minutes)
  A. General food consumption habits/practices and concerns
  B. Milk consumption
    Consumption frequency
    Reasons for drinking milk, types of milk purchased,
    Consumption patterns: When do you drink milk: breakfast/morning snack/
      lunch/afternoon snack/supper, evening or midnight snack?
  C. Milk purchase patterns:
    Where do you purchase milk (supermarket/grocery, convenience store,
      vending machine, restaurant, etc.)?
    Preferences for package size (gallon, half-gallon, pint, 8 ounce), purchase
      frequency, reasons for preference,
    Probe issues: Brand loyalty, quality of different brands of milk
    Probe issues: Factors influencing milk purchase (price, brand, container,
      etc.)
    Probe issue: Shelf-dating
  D. Shelf-life of milk
    How long does a container of milk typically last in your refrigerator? When
      you answer, tell us what size is that container.
    Have you ever discarded milk? Why?
    Probe: Packaging sizes available in the marketplace
    Do you discard milk that has passed the date stamped on the container?
    Probe: How long do you expect milk to last? Would you like it to last
      longer? Does it bother you to discard the milk? Would you rather run out
      of milk than have the milk go bad?

continued

Table 5.3   *Continued.*

  E. Milk quality
     How do you know when milk is no longer good to drink (i.e., smell, taste)?
     For those of you drink different types of milk such as 2% or any other type,
        do you notice any difference in storage life?
  F. Milk handling
     How do you handle milk? When you drink or use milk, do you leave the
        container on the table or counter for some time, or do you take the
        container out of the refrigerator for the shortest time possible?
  G. Milk flavor
     Have you ever detected any off-flavors or off-odors in milk stored in the
        refrigerator? (If spoilage has been mentioned in previous answers, then
        confine this question to off-flavors and off-odors other than spoilage.)
     Describe these off-flavors or off-odors.
     Probe: stale flavor or other off-flavors. How objectionable are these to you?
III. IN-DEPTH INVESTIGATION: Packaging aspects (60 minutes)
  A. Milk packaging
     Types and sizes of milk packages purchased
     Ask preferences for milk containers
  B. Opening and closure of milk packages
  C. Dispensing
     Difficulties in handling milk packages and dispensing milk
     Pouring milk from packages
  D. Carrying handles
  E. Refrigerated storage
     Probe: alternative shapes to improve space utilization
  F. Translucent vs. opaque packages
     Probe: Is it important to see milk through the package? Why?
     What colors so you prefer and which colors will you object to?
  G. Disposability of empty milk packages
     Probe Issues: Disposal of empty milk containers, recycling, ease of disposal
        of plastic jugs and paperboard containers?
  F. Alternative packaging
     Probe Issue: a new approach to milk packaging.
  G. Flexible pouches
     Probe: How do you feel about buying milk packaged in plastic pouches?
     Design of pouches
     Flexible or rigid sides, pouring from the rigid container or transfer to a rigid
        holder
     Dispensing spout, location on package, flush or protruding
  H. Design of multiunit pack
     Probe: Perceived refrigerated shelf-life
     Size of individual units and total volume of multiunit package
     Willingness to pay extra for the multiunit pack? How much extra?
     Probe: Suggestions for types of packaging material for outer container of the
        multiunit pack?

continued

Table 5.3 *Continued.*

I. Size of package and keeping quality of milk.
J. Labeling and point-of-purchase information.
   Probe issues: Milk labels, type size, color coding, recyclability, influence on
   purchase
IV. CLOSURE (10 minutes)
   Close, precede by false close.
   Thank you
   Distribute incentives

A common practice is to run at least three focus groups (Casey and Krueger, 1994). This practice may be based on finding out whether there are groups with divergent responses. However, because focus groups are qualitative and there is no need for quantitative projections to be made, three focus groups may not be necessary. The difference in opinions is an important result and should be reported. It may have implications for the clients, as in the need to develop different prototypes for different consumer groups.

Time management is important during the focus group. It is not advisable to keep participants beyond the two hours of participation to which they agreed.

Analysis and Interpretation

One of the most important tasks of the project leader is to discourage selective listening and caution against early conclusions and reports to management, before the results can be analyzed and reported.

Reporting

The report should be written as soon as possible following completion of the focus group. If written as soon as possible, it will be fresh in one's mind that the timeliness of the report will make it more likely to be acted upon. It helps to resolve any differences in interpretation between moderator and observers. To minimize turn-around time for the report, the objectives, and methodology sections should be written prior to conducting the focus group.

As with any qualitative research study, the results of focus group sessions are not intended to be generalized to a larger population (Fern, 1982). The analysis of data from focus groups should be truly qualitative. The counting of responses and tabulation of frequencies of responses serve little purpose. Qualitative conclusions must be objectively drawn. Quantitative generalizations from focus group results would be invalid, because the sample size and selection are limited. Nevertheless, qualitative research through focus groups can uncover attitudes and opinions found in the general population (Churchill, 1991). It is helpful to have a neutral, independent observer observe the sessions, transcribe the tapes, interpret the results, and write the report. The moderator may provide input based on observations and impressions during the discussion, but the moderator is usually preoccupied with eliciting responses from all panelists and making sure that all-important issues are covered in the time allotted for the focus group.

The report should contain an executive summary, background information, project objectives and methodology, including the number of groups conducted, the number of participants in each group, criteria for selection of participants, locations of tests, sample preparation, and presentation. A section of the report should caution the client on the limitations of qualitative research. Conclusions based on the objectives should be clearly stated. Recommendations should be made.

Handling Special Problems in a Focus Group

The information from the focus group is generated from the discussion. A common problem are those participants who are either overly talkative or do not respond to questions. The moderator should exhaust all means to encourage participation of all participants to ensure that all opinions about an issue are brought up during the discussion. The moderator may probe those lacking in consensus and encourage new opinions by asking if anyone disagrees with opinions that have been discussed.

Dominant Participants

A dominant respondent is described as one who may monopolize the discussion and, if allowed to do so, may influence the other participants. True experts can usually be screened out during recruitment. If not, participants may look to them to lead the responses. The other types of dominant participants are those that are self-appointed experts. When faced with this problem, the moderator can gain control of the situation by using nonverbal cues, such as avoiding eye contact with the dominant respondent, even looking at the ceiling or floor, angling his or her body away from the dominant participant and toward other participants, and directing questions directly to other participants. Other strategies are to use nonverbal cues such as drumming the fingers, stuffing notes or reading the Moderator's Guide, getting up and doing something in the room, or standing behind the dominant participant. If necessary, interrupt the dominant participant and redirect the question by giving permission for other views on the subject from other participants or change seating assignments so that the participants will be seated where they will face the same direction as the moderator rather than opposite from the moderator, where they feel they can challenge him or her.

Nonparticipants

This problem is caused by a respondent who does not talk or interact during the group discussions. The small number of panelists used requires that all participants be actively involved in the discussions. The moderator should try to get the panelist to contribute by facing and making eye contact with the nonparticipant, and directing questions to the nonparticipant such as, "John, how do you feel about the color of this beverage?" The moderator should nod or smile when a nonparticipant speaks or lean forward to show interest and encourage participation.

Unqualified Respondents

When it becomes obvious that a participant is unqualified, the moderator should use judgment. If the moderator realizes this at the beginning of the session, it is best to thank the person for showing up but to excuse the person from the focus group. If discussions are well under way, it may be best not to disrupt the discussion by excluding an unqualified participant. However, the moderator must take note of this during the analysis of results.

Low-Energy Groups

The moderator should increase the pace of the discussion by spending less time on each area before moving along to the next subject.

Legal Issues

It is important that all legal issues and documents be reviewed by a qualified legal authority and contain valid legal signatures before the test is conducted. This is to provide legal protection for the consumer agency conducting the survey, panelists, and the manufacturer of the product.

Misuse of Focus Group Results

The quantification of focus group results is dangerous. However, if it is used primarily as a method of generating hypotheses (ASTM, 1979), it has definitive value in the development process. The results are affected by factors arising from the group dynamics. Sometimes, even the most skilled panel leader may be unable to prevent biases that may arise.

References

ASTM, Committee E-18. 1979. *ASTM Manual on Consumer Sensory Evaluation*, ASTM Special Technical Publication 682, E. E. Schaefer, ed. American Society for Testing and Materials, Philadelphia, PA, pp. 5, 28–30.

Casey, M. A., and Krueger, R. A. 1994. Focus group interviewing. *In Measurement of Food Preferences*, H. J. H. MacFie and D. M. H. Thomson, eds. Blackie Academic and Professional, London, pp. 77–96.

Chambers, E., IV, and Smith, E. A. 1991. The use of qualitative research in product research and development. In *Sensory Science Theory and Applications in Foods*. H. T. Lawless and B. P. Klein, eds. Marcel Dekker, New York, Basel, and Hong Kong. pp. 395–412.

Churchill, G. A., Jr. 1991. *Marketing Research Methodological Foundations*. Dryden Press, Chicago, IL.

Fern, E. F. 1982. The use of focus groups for idea generation: The effects of group size, acquaintanceship, and moderator on response quantity and quality. *J. Market Res.* 19(2):1–13.

Galvez, F. C. F., and Resurreccion, A. V. A. 1992. Reliability of the focus group technique in determining the quality characteristics of mungbean (*Vigna radiata* (L.) Wilczec) noodles. *J. Sens. Stud.* 7:315–326.

Greenbaum, T. L. 1988. *The Practical Handbook and Guide to Focus Group Research*. Lexington Books, Lexington, MA.

Hashim, I. B., Resurreccion, A. V. A., and McWatters, K. H. 1995. Consumer acceptance of irradiated poultry. *Poultry Sci.* 74:1287–1294.

Hashim, I. B., Resurreccion, A. V. A., and McWatters, K. H. 1996. Consumer attitudes toward irradiated poultry. *Food Technol.* 50(3):77–80.

Kennedy, F. 1976. The focused group interview and moderator bias. Market. Rev. 31:19–21.

Lawless, H. T., and Heymann, H. 1997. *Sensory Evaluation of Food: Principles and Practices*. Chapman and Hall, New York.

Schutz, H. G. 1983. Multiple regression approach to optimization. *Food Technol.* 37(11):46–48, 62.

Sokolow, H. 1988. Qualitative methods for language development. In *Applied Sensory Analysis of Foods*, Vol. I., H. Moskowitz ed., pp. 3–19. CRC Press, Boca Raton, FL.

Teague, J. L., and Anderson, D. W. 1995. Consumer preferences for safe handling labels on meat and poultry. *J. Consum. Aff.* 29(1):108–127.

Wu, L. S. 1989. *Product Testing Consumer for Research Guidance*. ASTM, Philadelphia, PA.

6

Quantitative Methods by Test Location— Sensory Laboratory Tests

Introduction

Sensory testing requires special controls. If these controls are not adopted, results may be biased (ASTM, 1996). The commonly used method is the consumer acceptance test conducted in the sensory laboratory of the food company, consulting firm, or research organization. The laboratory test is the type of test that enables the researcher to exercise the most control over the testing and sample-preparation environment. These environmental controls include odor and lighting conditions, psychological distractions, and a comfortable testing environment.

Advantages and Disadvantages

Advantages

The major advantage of using a sensory laboratory to conduct a consumer affective test is the convenience of the location to the researchers. It is particularly appealing because of the accessibility of the laboratory to employees, and employees can be recruited to participate on short notice. The sensory laboratory is usually located within the building of the food company, consulting firm, or research organization. Although the sensory laboratory provides a convenient location for the research team and accessibility to employees, it is less convenient for local residents or other consumer panelists.

In instances when the panelists used are employees of the company, and if the employees have participated previously in consumer tests, they would be familiar with the testing procedures, thus saving the researcher time.

Among the other consumer tests, the laboratory test allows for the greatest control over the sample preparation and testing conditions, including lighting and environmental conditions. In the sensory laboratory, the researchers are able to control all conditions of the test such as product preparation and the product evaluation environment. Control of the product evaluation environment includes control of the testing environment such as lighting, noise, and other distractions, and conduct of the test in individual, partitioned booths, where panelists are isolated from each other. Lighting can mask color and other appearance factors so that panelists can focus on acceptance or other attributes under consideration.

Furthermore, because the sensory laboratory is usually adjacent to a fully equipped kitchen, sample-preparation steps can be standardized, including recipe formulation, the duration and temperature of cooking, holding and reheating, slicing (portion sizes of serving), and serving (with bread or crackers, dishes on which food is served), and can be conducted under highly controlled conditions.

One of the advantages of conducting a consumer test in the sensory laboratory is the rapid turn-around time for results to be obtained because of the proximity to the data processing facilities. When a computerized sensory data entry and analysis system is employed, feedback of results is more rapid.

The test involves only a few respondents, and tests can be conducted with as few as 25 to 50 consumer panelists.

Disadvantages

The disadvantages of conducting consumer tests in the sensory laboratory are the limitations on the consumer sample being used. An example of the limited information that may be obtained from a laboratory test is the lack of normal consumption such as sip versus drinking from a full glass. These limited procedures may influence detection of positive or negative attributes (Stone and Sidel, 1993; Meilgaard et al., 1991). Furthermore, product performance, when the food is prepared or tested in the laboratory, may be different from that in home use. Another disadvantage is the limited amount of time that the consumer is exposed to the product.

When consumers are recruited from a database, mailing lists will need to be maintained. If consumers screened from a prerecruited consumer database are used, most of those who would be willing to participate are those whose homes or place of employment are situated close enough to the sensory laboratory. Another disadvantage is a source of bias when company employees who are familiar with the test product are used; thus, care must be taken to recruit only those employees who are not familiar with the production, testing, or marketing of the product. On the other hand, local residents who participate in the tests may be biased due to the belief that the products are originating from a specific company or plant.

Role of the Project Leader

The project leader is responsible for facets of the test from the planning stage to the reporting of final results to the client or requester. The project leader will need to plan the design of the experiment; define the consumer panel characteristics; coordinate panelist recruitment and screening, questionnaire design and development, preparation and presentation of samples, the test procedures, testing environment, and quality control of all aspects of the test, data collection; and process, report, and interpret results.

The project leader coordinates and organizes the study. The project leader needs to understand the purpose and the goals of the study as stated by the client or requester in order to thoroughly define the problem and identify critical objectives. The project leader develops the consumer laboratory test plan to be submitted to the client or requester that delivers the information needed, outlines assumptions expected from the client and information that can be delivered as a result of the study. During the planning stages of the consumer laboratory test, the project leader must very specifically and clearly state the requirements of the test.

The project leader has several responsibilities during the planning and conduct of a consumer laboratory test. The major areas of responsibility for the project leader are as follows:

1. Prepare a clear statement of objectives.
2. Design the test, including a brief description of the test procedure sequence and schedule.
3. Describe in detail the test procedure, sample coding procedures, and methods to be used, including preparation of samples, suggested containers, serving procedures, serving instructions, serving size, timing, and lighting.
4. Identify test samples, number of test products, and number of responses per sample.
5. Design the questionnaire and pretest as needed.
6. Describe the consumer panel, panel size, and classification quotas for recruitment.
7. List screening criteria, prepare recruitment script, screener and monitor recruitment performance.
8. Specify requirements for booth area, including special lighting and other special needs.
9. List food product samples, ingredients, and other items to be purchased.
10. Assign personnel for recruitment, preparation, and serving of samples,

and other tasks needed for the test; identify additional personnel needs and their responsibilities.

11. Be responsible for planning and implementation of the study and quality assurance to ensure that correct testing procedures are followed.
12. Describe methods for data processing, analysis, and interpretation.
13. Decide on panelist incentives.

It is important for the project leader to monitor the performance of the agency and its personnel during all phases of testing. The project leader must ensure that the test protocol is implemented as outlined, the panelists are selected and screened as specified in the project plan, completed questionnaires are reviewed for completeness of responses, project activities are completed according to schedule, and the quality of the test is maintained throughout the test. For example, assigned samples are served in the appropriate sequence at the appropriate time, plate waste is monitored to ensure that enough sample is being consumed to make a valid evaluation, and the test samples are monitored for consistency throughout the study.

The necessity for sample consistency to be maintained is especially important if testing is to be carried out over a number of days. Completed questionnaires should be examined, especially at the beginning of the test, so that problems can be corrected early. After this, completed questionnaires should be examined periodically. Interviews should be closely monitored to prevent interviewer bias. It should be emphasized that any modifications or deviations from the defined test protocol must be approved in writing by the project leader. A checklist (see Appendix A) should be prepared to ensure that all tasks are completed in the time required for the test. The checklist will help the project leader keep track of the completion of tasks and increase the probability that the test will proceed smoothly.

Dry-Run and Briefing

It is recommended that a complete dry-run of all testing procedures be conducted on the test date, from the preparation of samples and orientation of panelists, to actual test procedures conducted using two or three untrained individuals as panelists one week before the test date. This will allow sufficient time for changes to be made if necessary.

The Consumer Panel

Selection of the panel, recruitment, and screening are discussed thoroughly in Chapter 4.

Panel Size

The consumer panel consists of consumers who are recruited and screened for eligibility to participate in the tests from a consumer database consisting of prerecruited consumers or company employees. Usually 25 to 50 responses are obtained; at least 40 responses per product are recommended by Stone and Sidel (1993), however, 50 to 100 responses are considered desirable (IFT/SED, 1981). In a test consisting of 24 panelists, it may be difficult to establish a statistically significant difference in a test with the small number of panelists. However, it is still possible to identify trends and to provide direction to the requestor. With 50 panelists, statistical significance increases to a large extent.

Sidel and Stone (1976) provided a guide for selecting panel size based on the expected degree of difference in ratings from the 9-point hedonic scale and the size of the standard deviations. They warn their readers that different products have their own requirements, indicating that smaller panels can provide statistical significance when the variability of the sample is small, and larger panels are needed as variability of the products increase.

Recruitment of Panelists

Preference or acceptance tests require different selection criteria for panelists compared to those in discrimination or descriptive tests. The makeup of the panel must be defined. In acceptance tests, panelists should be representative of the target market for the product.

One approach to recruitment is to develop a database of persons who may be available for testing. In many cases, demographic characteristics of the individuals in the database are known and updated on a regular basis. Recruitment of respondents may be done through a market research agency or by designated recruiters in the project team who recruit by telephone calls, referral, or personal contacts. Other methods often used to recruit panelists are random selection from a telephone directory, random-digit dialing, posters in retail stores, mailing lists from organizations, assorted consumer databases, and intercepts at malls, shopping areas, or restaurants.

Employees versus Nonemployees

Employees should not be used in affective tests unless sufficient tests have been conducted to ensure that the employees exhibit food consumption and preference patterns similar to that of the target market for the product.

Local Residents

One method of recruitment is to bring local residents into the laboratory. This approach is growing in use because it is convenient for the sensory staff in terms of scheduling of tests, the speed at which tests can be satisfied, and it minimizes the reliance on employees. However, the method of recruiting local residents involves prerecruitment, database management, scheduling, budgeting, accessibility, and security. A system must be established to contact and schedule local residents for a test. A budget and means of providing incentives or honoraria to panelists are required, which are not needed in tests involving employees. Accessibility of the laboratory to panelists so that they do not wanted in restricted areas is necessary. The advantages of using local residents is that participants are usually highly motivated, will show up promptly for each test, and are willing to provide considerable product information.

Another possibility is for a company to develop and maintain an off-premises test facility. This eliminates many of the problems associated with bringing local residents to the company premises. It will, however, be more costly than bringing people into the sensory laboratory. Another alternative is to contract with a sensory evaluation company, assuming they have the necessary sensory test resources.

In each case, the client will need to assess its current acceptance test resources and anticipated workload before determining which of these options are feasible. There is no question that consumer acceptance tests need to be done. The question is where they will be done, who will be the subjects, and the relative costs (Stone and Sidel, 1993).

Screening

For a given test, panelists need to qualify according to predetermined criteria that describe the target market for the product. A screener should be developed collaboratively with the sponsor of the test. The screener should be prepared so that it ensures reliability of the screening process and provides a tally to verify quotas. Selection of participants should be conducted as rigorously as possible. The undesirable practice of using those participants who can be most conveniently contacted, such as those people who live close to the facility, is often utilized but should be discouraged. Similarly, the use of friends and relatives of project staff should be avoided. These individuals may bias the results of the discussion and should not be used. Demographic criteria such as age, gender, frequency of product use, availability during the test date, and other criteria such as employment with the sponsoring company or similar business concerns, and other security screening criteria, are often used. In addition, ethnic or cultural background, occupation, education, family income, experience, and the last date of participation in a consumer affective test may be used. Individuals who have participated in several tests may no longer be naive consumers. Panelists with food allergies do not qualify as members of the panel. All persons who have in-depth knowledge of the product, or those who have specific knowledge of the samples and the variables being tested, should not be included in the test.

Attendance

The problem of "no-shows" should be avoided. The reduced panel size could be a source of embarrassment with clients. To minimize the attendance problem, a number of steps can be taken. It may be necessary to overbook according to a predetermined "no-show" rate. A rate of 20% may be high for a panel of consumers recruited from a database; a higher rate, possibly close to 50% or higher, would be expected when the panel is prerecruited through store intercepts. Therefore, to ensure participation, select participants who live approximately no more than thirty minutes from the sensory laboratory facility and give participants clear directions and a map to help them find the location; plan test dates so as not be in conflict with major community or school events. Send reminder letters or brightly colored postcards to post in a visible location, pay adequate fees or incentives, overbook, and, most important, make participants understand the importance of their attendance, promptness, and the value of their participation.

Incentives

Participants in laboratory tests are usually paid for their effort. The amount of incentive paid to participants may differ according to many factors, including length of the test, location of the test and associated travel expenses, and incidence rates of qualified participants.

Payment is given after the session is completed. Certain organizations may have restrictions on the type of incentive that may be awarded to panelists; therefore, these vary considerably. Incentives for participation may be provided in the form of a cash honorarium, selection from a gift catalog, gift certificates and tickets to special functions such as ball games or concerts, or donation to charity or a nonprofit organization.

Physical Facilities

Location of the Facility

The location of the sensory laboratory is important, because location may determine how accessible it is to panelists. The laboratory should be located so that it is convenient for the majority of the test respondents. Laboratories that are not convenient to go to will not only reduce the number of consumers who will want to participate but also limit the type of panelists that can be recruited for the tests.

It is best to locate the laboratory away from heavy traffic areas to avoid confusion and noise. For example, when the sensory laboratory is located in a company facility, preferably, it should not be situated near a noisy hallway, lobby, or cafeteria because of the possibility of disturbance during the test (ASTM, 1996). The panelists should not be able to hear the telephones and other office, food production, or laboratory equipment. If the sensory laboratory is located in those areas to increase accessibility to panelists, the laboratory should be equipped with special soundproofing features.

Sensory Laboratory Facility

The laboratory should be planned for efficient physical operation; furthermore, the laboratory should provide for a minimum amount of distraction of a panelists from laboratory equipment and personnel, and between panelists themselves (ASTM, 1996). The laboratory should be composed of separate food preparation and testing areas. These areas must be adequately separated to minimize interference during testing due to food preparation operations. The sample preparation should not be visible to the panel. Individual, partitioned booths are essential to avoid distraction between panelists; however, these should be designed so that they do not elicit the feeling of isolation from the rest of the panel.

It is important to have a reception area, separate from the testing and food preparation areas, where panelists can register and fill out demographic, honorarium, and consent forms, and wait before or after a test, without disturbing those who are doing the test. This area will allow room to encourage social interaction, to orient panelists on the test procedures, and to allow for payment of stipends.

Odors

The test area must be kept as free from odors as possible. A slight positive pressure in the testing room will reduce the flow of air from the food preparation areas to the booths. Air from the food preparation areas should be vented to the outside. Air intake from outside the building must not come from areas such as manufacturing exhaust vents or garbage dumpsters (ASTM, 1996).

Lighting

Adequate illumination is required in the evaluation areas for reading, writing and, examination and evaluation of food samples. Special light effects may be used to emphasize or hide irrelevant differences in color and other aspects of appearance. To emphasize color differences, different techniques such as the use of spotlights, changes in the spectral illumination by changing the source of light from incandescent to fluorescent, changing the types of fluorescent bulbs used, or changing the position of distance of the light source are used. To deemphasize or hide differences, a very low level of illumination may be used. Special lighting such as sodium lights, colored bulbs, or using color filters over standard lights may likewise be used. The color of the light may help reduce differences in appearance caused by hue (such as red or amber) but may do little to mask appearance differences such as degree of brownness or uniformity of color, or surface appearance (ASTM, 1996).

Testing Environment

The evaluation rooms must be comfortable enough to encourage panelists to concentrate on the sensory test. The temperature and humidity must be controlled to provide and maintain a comfortable environment. The furnishings must be comfortable and functional; therefore, all chairs and stools must be selected with care. The booth areas should be designed for comfort and concentration during testing.

Reception Room

The panel participants usually do not go directly to the booth area. They should be received in an area where they can hang up their coats, fill out necessary forms, and be comfortable until the orientation period.

Other Considerations

Orientation of Panelists

The orientation of consumer panelists should consist only of describing the mechanics of the test that they need to know. Examples of such topics are orientation regarding the booth area, which may include explanations about the sample pass-through door, signal lights, and so on. Orientation regarding the use of computerized ballots, such as instructions on using a light pen if these are used, is warranted. The orientation must be carefully panned to avoid any opportunity for altering the panelists' attitudes toward any of the food samples to be evaluated. Avoid giving any hint of the expected results of an experiment, and do not discuss the samples with the panelists prior to the test.

Preparation and Presentation of Samples

A detailed discussion of the different test methods used in affective testing, data collection, and questionnaire design and development is found in Chapter

2. A thorough explanation of test procedures, from the planning stage to data analysis procedures is in Chapter 3.

Test Procedures

The project leader will need to plan for preparation and presentation of samples, duration and temperature of holding after cooking until serving, portion size of serving, and the method of serving. The use of a carrier, dishes on which food is served, and whether or not to include bread, crackers or water for rinsing should be considered. The control of testing environment (red light, etc.) should likewise be considered.

Product Number per Sitting

For tests involving employees who cannot spend too much time away from their work assignment, the recommended number of products per sitting is two to six products, and questions should be limited to acceptance questions. Nonemployees are available for longer periods of time and can evaluate more products. The number of products that can be evaluated depends on the onset of sensory fatigue and the amount of sample preparation and handling required.

More samples can be tested if a controlled time interval between products is allowed and only acceptance is measured. Kamen et al. (1969) observed that under such conditions, consumers can evaluate as many as twelve products.

The use of complete block designs is recommended, even if panelists have to return to complete the test samples, being able to test a maximum of six samples per session. If this is not feasible, incomplete block designs may be used, but it is recommended (Stone and Sidel, 1993) that incomplete cells per panelist be kept to a minimum of one-third or less of the total number of samples being evaluated by the panelist.

References

ASTM, Committee E-18. 1996. *Sensory Testing Methods*. ASTM Manual Series: MNL 26, 2nd ed. E. Chambers, IV and M. B. Wolf, eds. American Society for Testing and Materials, West Conshohocken, PA. pp. 3–24.

Kamen, J. M., Peryam, D. R., Peryam, D. B., and Kroll, B. J. 1969. Hedonic differences as a function of number of samples evaluated. *J. Food Sci.* 34:475–479.

IFT/SED. 1981. Sensory evaluation guideline for testing food and beverage products. *Food Technol.* 35(11):50–59.

Meilgaard, M., Civille, G. V., and Carr, B. T. 1991. *Sensory Evaluation Techniques*, 2nd ed. CRC Press, Boca Raton, FL.

Sidel, J. L., and Stone, H. 1976. Experimental design and analysis of sensory tests. *Food Technol.* 30(11):32–38.

Stone, H., and Sidel, J. L. 1993. *Sensory Evaluation Practices*, 2nd ed. Academic Press, San Diego, CA.

7

Quantitative Methods by Test Location— Central Location Tests

Introduction

One of the important tools in obtaining guidance for maximizing product acceptance is the central location test (CLT). The CLT format may be adapted to use various psychophysical methodologies and to satisfy many different project objectives (ASTM, 1979).

Central location tests (CLT) are the most frequently used consumer tests, especially by those conducting market research. CLTs include a variety of techniques and procedures. The tests are conducted in one or, more often, several locations away from the sensory laboratory and are accessible to the public. These tests are usually conducted in a shopping mall or grocery store, school, church or hotel, or food service establishment, including cafeterias or similar types of location accessible to large numbers of consumers.

The psychological environment can be controlled, including such things as prior instructions to the subjects, manner of examining samples, freedom from outside interference, and ways of responding. It is recognized that these controls create some degree of artificiality. The subject's exposure to the product is necessarily very limited; as a result, his or her perception of the product may not be identical if used under conditions more closely approximating those of normal consumption. The panelist can only concentrate on a few features. Certainly, this would be the implication of the increasing popularity of central location testing. Product security is also an advantage, since the product does not leave the supervision of the company.

Advantages and Disadvantages

Advantages

The major advantage of the CLT is the capability to recruit a large number of participants and obtain a large number of responses. In addition, only actual consumers of the products may be recruited to participate in the tests; no company employees are used. The test will result in considerable impact and validity, because actual consumers are used. This type of testing enables the collection of information from groups of consumers under conditions of reasonably good control (ASTM, 1979) compared to a home-use test. Furthermore, several more products may be tested than would be advisable in a home-use test.

Disadvantages

Limited Resources

A disadvantage of the CLT is the distance of the test site from the food company, consulting firm, or research organization. Often, the central location has limited facilities, equipment, and resources necessary for food preparation and conduct of the test. Food preparation and testing facilities that may be lacking in a central location are space for a registration area, suitable sample-preparation areas, and individual partitioned booths for the sensory test. However, suitable equipment and testing space can be made available, and a small number of trained personnel may be used so that preparation of samples and serving of products can be controlled. Also, close supervision by the project leader can result in better control of stimulus characteristics compared to that in home-use tests.

Limited Controls and Types of Tasks to be Performed

The inadequacy of facilities would limit the types of tasks that can be performed by panelists and may pose a severe limitation on the ability to conduct the test under the experimentally controlled conditions required. The potential for distraction may be high. Panelists recruited for the test can readily walk out of the test area during a test. The large number of panelists that can be used for this type of test is as much a disadvantage as it is an advantage, because of the time and manpower requirements to conduct the test.

Limited Information

The products are tested under artificial-use conditions compared to normal use at home. Limited information is obtained as a result of the artificial conditions and the limited number of questions that can be asked. Limited information will be available from the data in regard to preferences of consumers from different age groups and socioeconomic groups if the test is limited to only one location or area.

Limited Exposure

The short time of exposure is at times listed as a disadvantage of both the CLT and laboratory tests compared to home-use tests, where the consumer will have more and longer exposure to a product (Stone and Sidel, 1993). Because both the CLT and laboratory tests are primarily for product screening and defining the consumer segment that will accept or prefer the product, longer exposure to the product is often unnecessary. Should the issue of acceptance, after a longer period of time in a consumer's home be important, a separate batch of samples should be subjected to conditions that simulate home storage (time, temperature, and relative humidity) and tested as a separate sample.

Role of the Project Leader

The project leader is responsible for the entire test, from the planning stage to the reporting of final results to the client or requester. The project leader is the coordinator and organizer of the study. It is essential that the project leader understand the purpose and the goals of the study as stated by the client or requester. The first step the project leader must take is to define thoroughly the problem and identify critical objectives of the test. The project leader develops a consumer test plan to be submitted to the client or requester that delivers the information needed by the client, and outlines assumptions expected from the client and information that can be delivered as a result of the study. During the planning stages of the CLT, the project leader must very specifically and clearly state the requirements of the test. This is especially true whether an agency implements the test or the test is performed by sensory project personnel.

The project leader has several responsibilities during the planning and conduct of a CLT. The major areas of responsibility for the project leader are as follows:

1. Write a clear statement of objectives.
2. Design the test, including a brief description of the test procedural sequence and schedule.
3. Describe, in detail, the test procedure, sample coding procedures, methods to be used, including preparation of samples, suggested containers and serving procedures, serving instructions, serving size, timing, and lighting.
4. Identify test samples, number of test products, and number of responses per sample.
5. Design the questionnaire and pretest as needed.
6. Describe consumer panel, panel size, and classification quotas for recruitment.
7. List screening criteria, prepare recruitment script, screener, and monitor recruitment performance.
8. State test site requirements, including special equipment needs.

9. List food products, ingredients, and other items to be purchased.
10. Identify personnel needs for recruitment, preparation and serving of samples, and other tasks needed for the test, and personnel responsibilities.
11. State cost constraints.
12. Identify special needs, and so forth.
13. Be responsible for planning and implementation of the study and quality assurance to ensure that correct testing procedures are followed.
14. Describe methods for data processing, analysis, and interpretation.
15. Provide suggestions for panelist incentives.

Whether an agency or project personnel is used to implement the test, it is important for the project leader to monitor the performance of the agency and its personnel during all phases of testing. The project leader must ensure that the test protocol is implemented as outlined, the panelists are selected and screened as specified in the project plan, interviews are conducted to strictly adhere to test protocol, interviewers do not introduce bias, completed questionnaires are reviewed for completeness of responses, project activities are completed according to schedule, and the quality of the test is maintained throughout the test. For example, assigned samples are served in the appropriate sequence at the appropriate time, plate waste is examined to ensure that enough sample is being consumed to make a valid evaluation, and the samples are monitored for consistency throughout the study.

The necessity for sample consistency to be maintained is especially important if testing is to be carried out over a number of days. Completed questionnaires should be examined especially at the beginning of the test so that problems can be corrected early. After this, completed questionnaires should be examined periodically. Interviews should be closely monitored to prevent interviewer bias. It should be emphasized that any modifications or deviations from the defined test protocol must be approved in writing by the project leader. A checklist (see Appendix B) should be prepared to ensure that all tasks are completed in the time required for the test. The checklist will help the project leader keep track of the completion of tasks and increase the probability that the test will proceed smoothly. Professionalism in style and dress by the project leader and agency personnel is an important consideration.

Panel

Panel Size

The number of consumers to be handled at one time depends on a number of factors that include the product type, the capacity of the testing facility, and the number of technicians available to conduct the test. One or two panelists at a time may be recruited to participate, or a group of panelists maybe handled at one time. The number of panelists will depend on the product type, the capacity of the testing facilities, and the number of personnel available to conduct the test. Too many panelists in one area will encourage inattention or interference, and if the panelist to technician ratio is too high, panelists' mistakes or questions may go unnoticed. In general, having more than twelve people at a time would not be optimum. Furthermore, the practice of administering a test simultaneously to a roomful of people, where the ratio of panelists to technicians is over one to twelve, would likely result in loss of control and would not be recommended.

Number of Panelists

Usually 100 (Stone and Sidel, 1993) or more consumers (responses per product) are obtained, but the number may range from 50 to 300 (Meilgaard et al., 1991), especially when segmentation is anticipated (Stone and Sidel, 1993). The increase in the number of consumers compared to the laboratory test is necessary to counterbalance the expected increase in variability due to the inexperience of the consumer participants and the "novelty of the situation" (Stone and Sidel, 1993). Several central locations may be used in the evaluation of a product. Tests that use "real" consumers have considerable face validity and credibility. The increased number of respondents to 100 or more has advantages and disadvantages compared to the laboratory test.

The number of products to be evaluated by a panelist per session at a CLT should be five to six, with fewer samples recommended. Consider carefully the number of samples that can be ideally presented at a mall location and at any test that employs the intercept method of recruiting panelists. In these instances, the panelists are more likely to walk out in the middle of the test if it is lengthy, complicated, or unpleasant.

The panelists in a CLT may consist of consumers who are recruited and screened for eligibility to participate in the tests from a consumer database of prerecruited consumers, or more often, they are intercepted at the central location where the test is to be conducted. The choice depends on the location of the test. If the test will be conducted in a hotel, school, or church, the panelists are often prerecruited using various methods. If the test will be conducted in a store or mall, consumers are more likely to be intercepted and asked if they would participate in the consumer test.

Prerecruitment requires more effort and time required by project personnel or an outside agency to contact eligible panelists, and this will have an additional, associated cost. With prerecruitment, there will be an additional cost of travel by participants to the test site. The price of prerecruitment exceeds that of intercepts, where panelists are usually given a small incentive such as product coupons.

Several factors need to be considered when planning for the central location site to be used and the type of recruitment that will be employed for the test. These include the frequency of use of the product or product category by consumers and the amount of sample preparation and product handling required for the test.

If a product has a low incidence of use (Stone and Sidel, 1993), such as in a product used by only 20% of the population that frequently goes to the central location, it would require an average of 500 intercepts to find 100 users of the product. Among these qualified users, two-thirds may be unwilling to participate in the tests for one reason or other. This brings up the number of intercepts required to get 100 test participants to 1,500 intercepts. Thus, when the incidence of use of the test sample is low, an extended period of testing will be required to complete the test. This can be remedied by conducting the test in several malls, but planning for such coupled with the potential increase in variability due to the different test locations, personnel, conditions of testing, and so on, presents additional challenges.

If the test product samples require extensive sample preparation and product handling, the benefits of the lower cost of mall intercepts may be diminished by the lack of food-preparation facilities, the increased time and cost of testing associated with the preparation of a large amount of sample, or its transport in the prepared state to the test site.

If prerecruitment is done, the recruitment will require time, which will differ according to the method used. However, once the participants are recruited, testing should be completed within one or two days.

Recruitment

The panel may be selected randomly or according to projected target-market specifications.

Intercepts

When the test is conducted at a store or shopping mall, consumer intercepts would be the best method of recruitment. Interviewers use visual screening for gender, approximate age, and race to select likely looking prospects from the traffic flow and ask the necessary screening questions. When a person qualifies as a panelist, the test is explained, and the person is immediately invited to come to the test center to take the test. Usually, a small reward in the form of money, coupons, or other payment is offered. This method of recruiting has become more popular and has been used more often during recent years.

Recruitment of Groups

Recruitment of panelists may be done through social or religious organizations such as clubs of different types, church groups, and chamber of commerce and school associations that receive a cash award for participation of their members. These groups provide homogeneous groups of participants and facilitate the recruitment process.

In these cases, the test is often conducted in facilities provided by the organization, such as a church or school hall, meeting room, and so on. The organization may either agree to provide a number of panelists of a certain type such as elderly consumers, homemakers with young children, or teens at intervals to fit the testing schedule, or provide the project personnel with lists of members who may be called, screened, and qualified ahead of time and scheduled.

A warning about the use of these groups is that they represent a narrow segment of the population (Stone and Sidel, 1993) and more often than not possess characteristics that are similar. In fact, many of them know or may be related to one another. Their responses may therefore be skewed.

Recruitment at Large Gatherings

One form of recruitment is to set up a booth or temporary facility at a convention, fair, industrial show, or similar event where crowds of people are likely to congregate. Visitors or passersby are invited to participate. This is usually limited to brief tests such as a "taste test" between two samples (ASTM, 1979).

Other Methods for Recruitment

Additional sources of panelists for CLTs are newspaper, radio, and television advertisements and flyers at community centers, grocery stores, and other business establishments, referrals from current panelists, letters to local businesses requesting their employees to become panelists, purchased mailing lists of consumers in a geographic location or telephone directories, and random-digit dialing.

Screening

The screening procedure needs to be unbiased. The selection of screening criteria is important, as results of the test can be meaningless if the proper target audience is not used. Thorough understanding of the study's objectives and discussions with the client or requester should have resulted in a screening questionnaire that will qualify only the desired participant. The actual screening criteria should be unknown so that consumers will answer with actual information about themselves, not what recruiters want to hear. The following issues are important and should be addressed when briefing the interviewers who will be implementing the study.

Location.   The location should be selected on the basis of matching the demographic profile of shoppers to the target consumer of the test product. Recruitment should be scheduled for the peak traffic hours for the desired participant. For example, if full-time employees are required, recruitment should be scheduled before or after regular working hours or during the lunch break; the cost of recruitment during regular working hours would be high.

Confidentiality and Security.   It may not be a wise practice to recruit panelists who are active panelists for another company. Individuals or immediate household members of individuals whose occupation is related to the manufacture or sales of the test product or in market research should be disqualified.

Past Participation.   These individuals may no longer perform as naive panelists depending on the degree or frequency of past participation in consumer tests. The frequency of participation of a panelist in tests within a product category may also be one of the screening criteria.

Other Factors.   Frequency of use of the product is important. Income level may be included as one of the criteria for screening potential panelists. Other factors such as age, marital status, gender, and education may be used as screening criteria. In some cases, the number of persons in a household, and whether these persons are adults, adolescents, or children, may be important. Whether an individual is the primary purchaser or preparer of food in the household may likewise be important. Finally, availability and interest of the panelist will determine whether he or she can be scheduled. Potential panelists who are allergic to the product should immediately be disqualified, as would individuals who are ill or pregnant.

Procedure

Any of the standard affective test methods can be used in CLTs. The decision on what test to use depends on the objectives, the product, and the panelists. Usually, the information is collected through a self-administered questionnaire.

In such cases, the questionnaire should be easy to understand and not take too long to answer. The attention span of the panelists must be taken into consideration. For example, if there are one or two samples, it is permissible to ask several questions about each; as the sample number increases, the number of questions that must be answered should be decreased accordingly. Panelists intercepted at a central location will often not agree to a test that will take over ten or fifteen minutes. Most panelists' interest and cooperation may be maintained for this length of time. Longer tests should be avoided unless there are special circumstances, such as giving the panelists a substantial cash incentive. Prerecruited panelists who are asked to come to a central location may be told beforehand how long the test will last. For example, in such cases, the panelists may be told that the test would take an hour.

On the test dates, panelists are assembled singly or in small groups in the test area where trained personnel conduct the test. The number of personnel will vary with the size of the group being handled, the stimulus control requirements in any given case, and the test procedure. For example, one-on-one interviewing may be required, and in other cases, it may be sufficient for one person to handle four to five people responding to a self-administered questionnaire at one time.

The panelists are given written or oral instructions to ensure adequate understanding of the test, including the number and type of test product samples to be tested, presentation of samples, waiting periods between samples, and other details of the test. The panelists are informed about the type of information to be collected by a brief review of the questionnaire.

Test Samples

The handling of test samples for a CLT is more involved because these have to be shipped from the laboratory to the test site. Sufficient samples for the test should be provided. Sample containers need to be clearly marked with appropriate sample-identifying codes and inventoried.

Sample preparation instructions need to be written in detail as clearly as possible and include holding time and temperature conditions. The timing of sample preparation is an important consideration. Timing must be planned so that sample preparation coincides with serving times. A dry-run is often necessary to ensure that the schedules are correctly timed. If preparation is going to be done in the test site, all necessary equipment for food preparation should be available. Temperature-holding devices such as warming lamps or steam tables should be available if needed. Training of personnel in sample preparation and equipment use may be necessary.

Sample serving specifications should be outlined. Appropriate serving utensils such as plates, plastic cups with or without lids, scoops, knives, or forks should be specified.

Test Facilities

Facilities for a CLT vary widely. In general, testing should be conducted at a location that will be convenient for the participants to travel to, with ample parking space. Access to public transportation may also be an important consideration.

Much of the testing may occur in settings that are less than optimal. Two general requirements for test facilities are agreed upon: (1) The facilities should allow adequate space and equipment for preparation of the product and for presentation of the product to panelists in a controlled environment; (2) the facilities should provide proper control of the "physiological and psychological test environment" and includes adequate lighting, temperature, and humidity control for the general comfort of panelists, freedom from distractions such as odors and noise, and elimination of interference from outsiders and between subjects (ASTM, 1979).

Central location testing can be contracted out to a marketing research company that operates its own testing facility. A CLT facility may be purchased, constructed, or leased. In such cases, proper control can be maintained in the test area by constructing a sample-preparation laboratory, panel booths, and adequate ventilation systems.

CLTs using groups are often conducted in public or private buildings such as churches, schools, firehouses, and so on. This type of test offers the least amount of control. Testing in these sites is usually done without booths, controlled lighting, or ventilation. Noise and odors may likewise pose a problem. Sample preparation must be done in advance, and test samples and equipment will need to be transported to the test facility. The problem of clean-up and waste disposal may involve hiring of a custodian when the test is conducted in these facilities.

Ideally, panelists should be in individual, partitioned booths while testing. Isolating the panelists eliminates distractions from other panelists or test personnel. When booths are not available, panelists should be isolated from each other as well as possible to minimize distractions and provide privacy while testing. Seating should be comfortable and at an appropriate height for the table or counter where testing will take place.

If using a test site for the first time, it should be examined at least one week prior to the test to be sure it will be sufficient. If the test site has been used previously, evaluation may take place a day or two before the test date. Final details may be arranged then.

Equipment should be tested to ensure it is in good working order. For example, ovens should be calibrated prior to the test day. Food samples, ingredients, and other supplies should be examined for quality and set up for the test. Storage areas and containers for samples and supplies requested in advance should be examined. Whenever possible, a dry-run of the sample preparation using the equipment should be made.

Samples should be prepared out of the sight of the panelists and served on uniform plates or sample cups and glasses. The serving area should be convenient to both the test personnel and the sample-preparation area. The serving area should have the serving scheme posted for test personnel to mark as they obtain samples for each panelist.

Large containers for waste disposal need to be positioned near the sample-preparation and serving areas. Arrangements need to be made for waste disposal from the test site after completion of the study. Samples that were not used should be handled according to the client's directions. Instructions regarding the handling of the product should be agreed upon beforehand.

Reception Area

Provisions must be made for panelists to have a comfortable place to wait once they arrive for the test. When planning for a reception area, sufficient space should be available for panelist registration, orientation and waiting for the test, payment of panelists, and a place for treats and social interaction after a test. In some cases, an area for child care is designated to allow panelists to take turns baby-sitting while they take turns testing. It may be helpful to establish rules regarding children and to post these rules so that fewer problems will arise.

Other Considerations

Dry-Run and Briefing

In a briefing session before the actual test, it is essential that the project leader review the complete test protocol to determine whether the test personnel are familiar with their assigned tasks, the procedures, and the serving instructions. It is a common practice for an agency to conduct a briefing with personnel prior to the test. The briefings are a good opportunity to review the instructions for the study and explain any special requirements. If a script will be used, a dry-run of the reading of the script is conducted to identify any problems. The importance of reading scripts verbatim is emphasized. The serving scheme for test samples is explained during this time.

References

ASTM, Committee E-18. 1979. *ASTM Manual on Consumer Sensory Evaluation*, ASTM Special Technical Publication 682, E. E. Schaefer, ed. American Society for Testing and Materials, Philadelphia, PA, pp. 28–30.

Meilgaard, M., Civille, G. V., and Carr, B. T. 1991. *Sensory Evaluation Techniques*, 2nd ed. CRC Press, Boca Raton, FL.

Stone, H., and Sidel, J. L. 1993. *Sensory Evaluation Practices*, 2nd ed. Academic Press, San Diego, CA.

8

Quantitative Methods by Test Location— Mobile Laboratory Tests

Introduction

A special form of test that can bring the advantages of both the laboratory test and the central location test in one test is the mobile laboratory test. Mobility brings with it a number of benefits (Mermelstein, 1988). The use of a mobile laboratory provides facilities for sample preparation and testing that can be environmentally controlled much like a permanent laboratory facility, yet provide the numbers and diversity of consumers that can be intercepted at a central location.

Procedure

Almost any type of affective test can be conducted in a mobile laboratory. Participants are usually recruited through intercepts made at the location where the mobile laboratory is parked. The test is usually run by sensory staff. Data are collected by trained interviewers. The recommended number of products to be evaluated is two to four and preferably not more than five. Because the panelists are inexperienced, tasks should be limited. Planning would be similar to that needed in arranging a central location test.

Advantages and Disadvantages

The advantages and disadvantages of using a mobile laboratory represent a combination of those listed for the laboratory test and the central location test.

Advantages

The major advantage of using the mobile laboratory for a consumer affective test is the ability to recruit a large number of "real" consumers to participate in the test and have the space to maintain experimentally controlled facilities and environmental conditions for food preparation and testing. It is convenient for the sensory personnel to schedule the test. The sample-preparation and testing facilities may allow better control of the testing than many central location test facilities have to offer. The mobile laboratory can be driven almost everywhere and can be parked anywhere there is a specific target population that can be intercepted for the test. The access to large numbers of consumers contributes to the short amount of time needed to obtain the required number of responses. No employees are used as participants in the test. Finally, because the tasks in a mobile laboratory test are limited and panelists do not have to travel to the test location as they would in tests held in a laboratory, panelists' incentives can be either minimal or unnecessary.

Disadvantages

The disadvantages of the mobile laboratory test are the expense of the purchase and maintenance of the laboratory. Logistical arrangements have to be made prior to the test for parking and use of power plug-ins. When conducting a mobile laboratory test, the planning and preparation phases are more rigorous compared to a laboratory test, where the facilities and equipment are located in the same building or within walking distance from the test site. Computerization of the mobile laboratory may impose additional electrical requirements. Furthermore, not all types of food can be tested in the mobile lab. Those food samples that require specialized equipment in their preparation will need to be prepared beforehand. The food-preparation equipment installed or used in a mobile laboratory is usually limited to kitchen-type equipment such as kitchen ranges, traditional and microwave ovens, refrigerators, and kitchen appliances.

Usually, a mobile laboratory test is best conducted using trained interviewers. Interviewers need to be trained to standardize collection of data and minimize bias during the interviews.

The mobile units usually range from vans and buses as short as fifteen feet long to tractor trailers eighty feet long (Mermelstein, 1988). The test conditions are limited by space constraints. Partitioned booths are likely to be unavailable, and there may be one or two testing stations.

Unless the space constraints allow for testing of several consumers at one time, the tests conducted in a mobile laboratory must require little time to complete; thus, only a small amount of information is obtained. The panelists can be given a limited number of instructions and, as in a central location test, no lengthy or unpleasant tasks are advisable or consumers may walk out. The potential for distraction and increased variability may be high. Consumers who have been intercepted and agree to participate in consumer tests in the mobile laboratory usually object to tests that take over fifteen minutes. This limits the number and complexity of tasks that can be conducted and contributes to the limited amount of information that can be obtained.

As in central location tests, limited information may be obtained as a result of using consumers from one area that may possess similar demographic and socioeconomic characteristics. This problem can be easily avoided when using a mobile laboratory by conducting the tests in several strategically selected locations over a period of days.

Role of the Project Leader

The project leader is responsible for the coordination and organization of the mobile laboratory test. The project leader needs to understand the purpose and the goals of the study as stated by the requester of the test. Then, the project leader must thoroughly define the problem and identify critical objectives. The project leader develops a project plan to be submitted to the requester or client that delivers the information needed by the client and outlines assumptions expected from the client and information that can be delivered as a result of the study. During the planning stages of the test, the project leader must very specifically and clearly state the requirements of the test.

The project leader has several responsibilities during the planning and conduct of the test. The major areas of responsibility for the project leader are to

1. Write a clear statement of objectives.
2. Design the experiment.
3. Describe in detail the test procedure, sample coding procedures, methods to be used, including preparation of samples, suggested containers and serving procedures, serving instructions, serving size, and timing.
4. Identify test samples, the number of test products, and the number of responses per sample.
5. Design the questionnaire and visual aids, such as enlarged scales to be used in the test to aid in collection of consumer responses, and pretest these as needed.
6. List screening criteria for consumer participants.
7. List special equipment needs, such as microwave ovens for reheating food samples and heating lamps to maintain product temperature after cooking or reheating.
8. List food products, ingredients, and other items to be purchased.
9. Identify personnel needs for recruitment, preparation, and serving of samples, and other tasks needed for the test; identify personnel responsibilities.

10. Identify special needs, such as prearrangement with shopping centers or grocery stores to park near the store entrance and use electric power plug-ins.
11. Be responsible for planning and implementation of the study, including interviewer training, dry-runs, and quality assurance to ensure that correct testing procedures are followed.
12. Describe methods for data processing, analysis, and interpretation.
13. Provide suggestions for panelist incentives.

It is important for the project leader to monitor the performance of the test personnel during all phases of testing. The project leader must ensure that the test protocol is implemented as outlined, and that interviews are conducted to strictly adhere to test protocol and do not introduce bias. Questionnaires completed by the trained interviewers are reviewed for completeness of responses, project activities are completed according to schedule, and the quality of the test is maintained throughout the test. For example, assigned samples are served in the appropriate sequence at the appropriate time, and the samples are monitored for consistency throughout the study. The necessity for sample consistency to be maintained is especially important if testing is to be carried out over a number of days.

Panel

The panel often consists of consumers who are intercepted to participate in the tests. The number of consumers to be tested at one time depends on a number of factors, including the type of food products to be tested, the capacity of the mobile laboratory testing stations, and the number of personnel available to conduct the test. One or two panelists may participate in the test, or if the mobile lab is large enough to handle more consumers, a larger number of interviews may take place simultaneously. There is usually no prerecruitment or screening except for age, gender, and food-product usage patterns.

Tests conducted in a mobile laboratory usually involve forty to sixty responses per product (Stone and Sidel, 1993). With minimal recruitment effort, seventy-five or more responses can be obtained in one location within a few hours. The mobile laboratory test is usually conducted in more than one location, and several central locations may be used in the evaluation of a product. For example, in acceptance tests on a fried product conducted by McWatters et al. (1990) in a mobile laboratory, consumers were intercepted and asked to participate. A total of 450 consumers participated, with 150 responses obtained at each of three locations.

Figure 8.1 Mobile Research Laboratory used in sensory affective tests.

Physical Facilities

A mobile sensory laboratory is a motor vehicle such as a large converted van, mobile home, or trailer built with facilities for food preparation and sensory testing. The mobile laboratory facilities can include a registration or receiving area, complete food-preparation facilities equipped with kitchen appliances, and small laboratory equipment, testing stations, lighting, and environmentally controlled conditions. The food-preparation equipment installed or used in a mobile laboratory is usually limited to kitchen-type equipment such as kitchen ranges, traditional and microwave ovens, refrigerators, and kitchen appliances. For example, in mobile laboratory tests conducted by McWatters et al. (1990), packages of the frozen product were transported in ice chests to the survey site, then heated in a microwave oven to an internal temperature of 70°C prior to serving. Drinking water for rinsing was offered to all panelists.

A mobile laboratory (Figure 8.1) that was used extensively in consumer tests (Resurreccion and Prussia, 1986; McWatters et al., 1990) was designed by Prussia and Tollner (1984). Many human-factor principles were used during all stages of designing and building the laboratory. The arrangement of the workspace incorporated a work triangle formed by the sink, stove, and refrigerator. A floor plan of the mobile laboratory is shown in Figure 8.2. The refrigerator

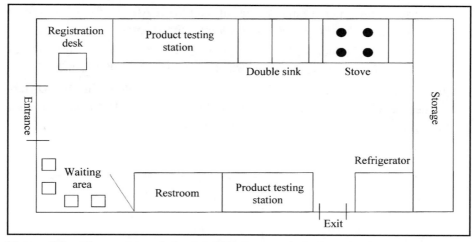

Figure 8.2 Floor plan of the Mobile Research Laboratory used in sensory affective tests. The entrance is at the back section, where registration is handles. The two product testing stations are offset to minimize congestion.

was located near the exit door to allow ready access from the outside. Other arrangement decisions taken into consideration were for two interviews to take place at any given time. The aisle width was adequate for a third worker to pass the two interviewers and a panelist at each station. The heating, cooling, and ventilating equipment provided a comfortable testing environment. Illumination and vibration levels were satisfactory, and the noise level was acceptable due to adequate isolation of mechanical equipment such as the electric generator sets, water pumps, and air conditioners.

The mobile laboratory can be driven to different locations where large numbers of people congregate, such as shopping centers, mall parking lots, school or church grounds, or fairs. The testing is done within the mobile laboratory using a method similar to that in any laboratory test, except that usually the participants of a mobile laboratory test are intercepted for the test.

References

Stone, H., and Sidel, J. L. 1993. *Sensory Evaluation Practices*, 2nd ed. Academic Press, San Diego, CA.

Mermelstein, N. H. 1988. Mobile units: Bringing technology to the user. *Food Technol.* 42(12):119–134.

McWatters, K. H., Resurreccion, A. V. A., and Fletcher, S. M. Response of

American consumers to akara, a traditional West African food made from cowpea paste. *Intl. J. Food Sci. Technol.* 25:551–557.

Prussia, S. E., and Tollner, E. W. 1984. Human factors for mobile agricultural research laboratory. Transactions of the ASAE. 27(4):997–1002.

Resurreccion, A. V. A., and Prussia, S. E. 1986. Food related attitudes: Differences between employed and non-employed women. *ACCI 32nd Annual Conference Proceedings.* K. P. Schnittgrund, ed. p. 156.

9

Quantitative Methods by Test Location— Home-Use Tests

Introduction

The home-use test (HUT) is also referred to as the home placement or in-home placement test. This test, as the name implies, requires that the research be conducted in the participants' own homes. It provides testing conditions that are not researcher-controlled and thereby could yield the most variable results. The HUT is used to assess product attributes, acceptance/preference, and performance under actual-use conditions. The food product samples are tested under normal-use conditions. They therefore provide information regarding the product that may not be obtained in any other type of test. From the product developer's point of view, they can provide information about the sensory characteristics of a product under uncontrolled conditions of preparation, serving, and evaluation.

HUTs are valuable in obtaining measurements about products that are difficult to obtain in a laboratory setting. An example of this is to determine consumer acceptance and use of irradiated chicken, wherein the researcher needs to know not only how much the consumer likes the irradiated chicken samples, but also the performance of the packaging and how the consumer will cook and serve irradiated chicken. Since products can fail due to packaging or product use, a HUT would be able to test the capability of the packaging early in the development phases of the product. Knowing how consumers will cook and serve irradiated chicken will provide information on product use by the consumer.

Advantages and Disadvantages

There are several advantages and disadvantages associated with HUTs.

Advantages

The major advantage of the HUT is that the products are tested in the actual home environment under actual, normal home-use conditions. Another advantage of conducting HUTs is that more information is available from this test method because one may obtain the responses of the entire household on usage of the product. Responses may be obtained not only from the respondent, who is usually the major shopper and purchaser of food in the household, but also from the other members of the entire household. This consumer test method can be used early in the product-formulation phase, where it is not only able to test a product for acceptance or preference but also for product performance. In addition, marketing information can be obtained, such as the types of competitive products found in the home during the test, usage patterns, and other information that would be useful in marketing the product. Information regarding repeat-purchase behavior may be obtained from the consumer (Hashim et al., 1996).

Participants may be selected to represent the target population. If the participants of the tests are prerecruited from an existing database and screened, the participants are aware of their role and the importance of the data collected, and this will likely result in a high response rate. When using a mail panel, speed, economy, and broad coverage are possible.

Disadvantages

The main disadvantages of the HUT are that it requires a considerable amount of time to implement, to distribute samples to participants, and collect participants' responses. It often takes at least one to four weeks to complete (Meilgaard et al., 1991). Lack of control is another disadvantage; little can be done to exert any control over the testing conditions once the product is in the home of the respondent. This lack of control may result in large variability of responses. The test design must be simple—it is best to conduct a HUT on only one or two samples; otherwise, the test situation would be too complex for most respondents. Therefore the HUT is not appropriate when conducting multisample tests. The HUT is the most expensive test in terms of product cost when test product sample sizes are larger than that served to panelists in a laboratory or central location test. In addition, mailing or distribution of the products to participants will add to the cost of the test. On the other hand, if the consumer panel size is smaller than that in a laboratory or central location test because of the cost, the information one would obtain from this test is limited. Furthermore, if using a mail panel, perishable and nonmailable products cannot be tested. If participants have not been prerecruited and screened from an existing database, the participants will likely be less aware of the importance of their role in this test, and the importance of the data being collected. In such cases, response rates will be lower than desired unless respondents are frequently reminded. Questionnaires must be clear and concise, because there is no opportunity to explain and elaborate as one would be able to do in a personal interview. Visual aids, if any are needed, will be limited to graphics on a questionnaire. Finally, the researcher needs to realize that the consumer respondent will be able to read all the items on the questionnaire beforehand, thereby limiting the impact of sequencing of questions.

Role of the Project Leader

The project leader is responsible for coordinating and organizing the study. The project leader needs to understand the purpose and the goals of the study as stated by the client or requester. The project leader must thoroughly define the problem and identify critical objectives. The project leader develops a project plan to be submitted to the client or requester that delivers the information needed by the client, outlines assumptions expected from the client and information that can be delivered as a result of the study. During the planning stages of the HUT, the project leader must very specifically and clearly state the requirements of the test. This is especially true whether an agency implements the test or the test is performed by sensory project personnel.

The project leader has several responsibilities during the planning and conduct of a HUT. The major areas of responsibility for the project leader are to

1. Write a clear statement of objectives.
2. Design the test.
3. Describe, in detail, the test procedure, methods to be used, including preparation of samples, sample size, suggested containers, sample coding procedures, and sample placement procedures.
4. Identify test samples, number of test products, and number of respondents.
5. Prepare instructions to consumers and design the questionnaire and pretest as needed.
6. Describe consumer panel, panel size, and classification quotas for recruitment, and recruitment method.
7. List screening criteria.
8. Identify personnel needs for recruitment of respondents, preparation, and packaging of samples, placement of samples, data collection, and other tasks needed for the test; identify personnel responsibilities.

9. Identify special needs, and so forth.
10. Be responsible for planning and implementation of the study and quality assurance to ensure that correct testing procedures are followed.
11. Describe methods for data processing, analysis, and interpretation.
12. Provide suggestions for panelist incentives.

Whether an agency or project personnel are used to implement the test, it is important for the project leader to monitor the performance of the agency and its personnel during all phases of testing. The project leader must ensure that the test protocol is implemented as outlined, the panelists are selected and screened as specified in the project plan, interviews are conducted to strictly adhere to test protocol, interviewers do not introduce bias, completed questionnaires are reviewed for completeness of responses, project activities are completed according to schedule, and the quality of the test is maintained throughout the test.

The necessity for sample consistency to be maintained is especially important if testing is to be carried out over a number of days. Completed questionnaires should be examined especially at the beginning of the test so that problems can be corrected early. After this, completed questionnaires should be examined periodically. Interviews should be closely monitored to prevent interviewer bias. It should be emphasized that any modifications or deviations from the defined test protocol must be approved in writing by the project leader. A checklist should be prepared to ensure that all tasks are completed in the time required for the test. The checklist will help the project leader keep track of the completion of tasks and increase the probability that the test will proceed smoothly. Professionalism in style and dress by the project leader and agency personnel is an important consideration.

Panel

Panel Size

All responses required in calculations for sample size should be obtained before the data are analyzed. Therefore, the sample size for the HUT should be large enough to get the necessary number of respondents within a reasonable length of time. Most of the time, mail panels are represented as being quota samples that match U.S. census distribution. Most agencies guarantee a 70% response rate. The factors of cooperation and nonresponse may influence the results of these samplings (ASTM, 1979). The assumptions are that the nonresponders will react similarly to those who respond—fortunately, there are instances where mail panel results closely approximate probability sampling results. The apparent superiority of sampling of respondents may be, at least partly, an illusion. People who volunteer for such activity and sustain their interest may be different from the normal user of the product type. Also, it is often alleged, "though seldom proven" (ASTM, 1979), that panel members who are used repeatedly may develop special ways of responding that are not typical of consumers.

Recruitment and Placement

In HUTs, product samples are tested in the consumers' own homes. The test involves preference, acceptance, performance (intensity and marketing information), and can also provide data on product preparation, attitudes, and other types of information.

There are various methods of locating and recruiting respondents, and delivering the test products. One method is to prerecruit from a consumer database and screen for panelists' eligibility to participate in the tests. However, the practice in some cases is to involve employees of a company who have little or no responsibility for production, testing, or marketing of a product. Due to the uncontrolled conditions of testing, a larger sample than that required for a laboratory test is recommended for a HUT. Usually, 50–100 responses are obtained per product. The number varies with the type of product being tested and with the experience of the respondent in participating in HUTs. With "test-wise" panelists, Stone and Sidel (1993) recommend that a reduction in panel size could be made, because these panelists who know how the test is conducted and feel more comfortable in the test situation will be less susceptible to error and the result of psychological variables associated with being a subject. In multicity tests, 75 to 300 responses are obtained per city in three or four cities (Meilgaard et al., 1991).

Another approach is to use the telephone for preliminary questioning to establish qualifications and solicit cooperation. Calling may be random within a given area or from lists of members provided by cooperating organizations. This approach has an advantage in that the recruiting is less expensive when qualified respondents are expected to occur at low frequency. On the other hand, it is probably harder to gain cooperation over the telephone, and the problem of delivering products and test instructions has to be solved. Furthermore, more effort may be required in recontacting the participants, because they may be more geographically dispersed.

Another approach is to survey door-to-door in areas selected on the basis of having a high probability of producing desired subjects. An interviewer asks questions to establish qualification and to obtain background information that might be useful in analyzing the results. When a qualified family is found, the placement is made on the spot, instructions are given, and arrangements are made for the return interview. An advantage of this approach is that it permits distributing the sample as desired in a given territory. Also, it eliminates the problem of product delivery, and people are easier to contact for the final interview. However, the method may be more expensive, because much effort is wasted in contacting unqualified people.

Mall intercept or in-store recruitment has been relied on to an increasing extent for many types of market research studies. Although mall intercepts are more commonly used in central location tests, they may be used for HUT recruitment. Interviewers recruit prospective participants in the mall or store and can quickly screen them to determine whether they meet requirements. If cooperation is solicited and agreement is reached, the test products and instructions can be supplied with little delay. At times, bulky or perishable products are delivered to the respondents' homes at a later date. The advantage of this approach is that it permits screening a large number of prospects relatively quickly and cheaply, which can be particularly helpful when trying to locate users of a low-incidence product type. A serious disadvantage of this recruitment method is that it may be time consuming and expensive to recontact the respondents, since their residences will most probably spread over a wide territory.

If panelists are to be reached by mail, a necessary step is the development of databases of mailing lists and keeping these current by rotation and replacement of dropouts. The development of a database involves initial location of the respondents, contact, gaining cooperation, acquiring necessary information, and maintaining sufficient interest for sustained participation. Contacts maybe made through the telephone through random-digit dialing, newspaper advertisements, and recruiting through organizations, and so on. Prospective panel members provide information about themselves, which is placed on file and retained for a certain period of time. The frequency with which a panel member may be contacted varies widely. Usually, they are given some incentive for their participation.

The key to obtaining a valid sample of respondents is the initial selection and gathering of data. The usual approach is to seek protection in large numbers, with lists containing thousands of names and representing more or less the national census distribution of the population. Information is obtained by means of questionnaires. Such demographic data as location, the number and identity of family members, family income, race, education, and occupation are nearly always available. Beyond this, it is a matter of product usage.

This information is stored in a computer. Drawing the actual sample for a given study is a mechanical matter of identifying families with the desired characteristics, then randomly selecting the desired number. When a HUT is being planned, a sample of respondents that meets the requirements of the test is drawn from the database.

Facilities

HUTs are conducted in the respondents' own homes. No controls of the environment or the testing conditions can be made.

Procedure

The test is conducted in a location wherein the researcher has no control over the testing environment. The products are tested under actual-use conditions, and responses of one or all family members may be asked for. Therefore, the test design should be as easy to perform and uncomplicated as possible. The focus should be on consumer acceptance, and the test can provide the researcher with product-use variables that may be determinants of product acceptance by the consumer.

The test requires additional decisions that need to be resolved prior to the test. These include containers and labels, product-preparation instructions that are not ambiguous and must be complete, product placement—whether mailed or delivered to consumers' homes—and the self-administered HUT questionnaire.

Samples

The product containers to be used in HUTs are often those in which the product will be sold but should be plain and have no graphics. Containers should be labeled with the sample code number and a label containing the preparation instructions, contents, and a telephone number to call in case questions arise.

Preparation instructions must be complete and easily understood. These must describe exactly how the products are to be prepared. Directions on completing the questionnaire should be included. The instructions should be unambiguous and be prepared so that they are clear to the panelist and do not cause confusion. Pretesting the instructions and the questionnaire is important to the success of a HUT.

Product Placement

The number of products to be tested should be limited to two, primarily because of the length of time needed to evaluate each product, which may last over four days to one week, and the scoresheet is rated, then the second product is supplied and rated. If additional products are included in the test, the test would take longer, and the risk of nonresponse would be greater due to one or more reasons such as loss of the questionnaire, trips out of town, or panelists' loss of interest.

There are three methods that can be used to place food product samples in consumer panelists' homes. These are to (1) mail the product, (2) deliver the product, or (3) have consumer panelists come to a central location to obtain the product. Providing the respondent with all the product during the beginning of the test minimizes cost. However, lower costs can be achieved by mailing products to the consumers prerecruited by telephone or mail to participate in the test. Mailing products to subjects is generally risky and should be avoided except in rare instances, such as when the products are microbiologically shelf-stable and are of a shape and size that lends well to mailing. An example of this is using the mail to place packaged pecan halves to participants of a HUT (Resurreccion and Heaton, 1987). The problems with using the mail to place products with participants is the time required, the choice of mailers, microbiological safety, and that the package may be received by someone other than for whom it was intended, resulting in missing data.

Providing two or more products at the same time is likewise not recommended for the following reasons: It allows the participants to make direct comparisons and increases the probability of writing responses for the wrong product on the questionnaire (Meilgaard et al., 1991) and invalidates results. For these reasons Stone and Sidel (1993) recommend that cost should not be an issue when deciding on placement of samples in a HUT.

The practice of having consumers recruited for the test to come to a central location to obtain product is efficient (Hashim et al., 1995). If there is a second sample, consumers will return to the location to return their empty container, and completed questionnaire in exchange for the next product, and a new questionnaire. This method is more effective and cost-effective than delivering the product to consumers' homes when local residents are used.

In certain cases when the method of sampling for a HUT is to intercept and qualify prospective participants, qualified participants may be given the first of the samples immediately after qualifying. They will then return to the same location to pick up the next sample and questionnaire, and return the empty sample container and questionnaire.

Measurement of Product Acceptance

The primary concern in a HUT is measuring overall acceptance of a product. To maintain uniformity and continuity (Stone and Sidel, 1993) between different methods of testing such as the laboratory test, central location test and HUT, the same acceptance scale should be used in the HUT. There are several practices that should be considered when designing the questionnaire:

1. Inclusion of attribute diagnostic questions.
2. Addition of product performance questions.
3. Paired-preference questions.

In addition to overall acceptance, acceptance of specific product attributes such as appearance, aroma, taste, and texture may be obtained, but the results should not be used as one would results from a sensory analytical test. Objections to using diagnostic questions were raised by Stone and Sidel (1993), who recommend that diagnostic questions be excluded from the test. One problem is the lack of assurance that the responses are valid. The second is that the overall acceptance requires the evaluation of the product as a whole, without direction as to what sensory aspects are considered important. On the other hand, diagnostic assessment directs subjects to focus on specific sensory components from an analytical perspective.

Often, there is interest in learning more than overall acceptance from a HUT. Questions such as product performance, "How easy is it to pour milk from this stand-up pouch?" or purchase intent, "If available in the market, how often would you purchase this product?" can be easily obtained. Although these questions are related to acceptance of the product, they do not require an analytical sensory set, and as such, should not create the same problems with attribute acceptance questions (Stone and Sidel, 1993).

Some researchers ask a paired-preference question on products that were tested during two different time periods. This final question, if added, relies on the consumers' memory to remember what was tested the week before and to compare it with the product currently being tested. This use of the paired-preference question is inappropriate. The 9-point hedonic scale for each product yields more useful information than the final paired-preference question.

Data Collection

There are various ways of obtaining the critical post-use information about the respondents preferences, attitudes, and opinions. These are the personal interview, use of a self-administered questionnaire that is mailed back, and the telephone interview.

Personal Interview

The most effective approach among the three methods is the personal interview. The personal interview allows the researcher to review procedures with the respondent and determine whether the product was used properly, and to clarify any questions regarding test procedures or product use conditions. Respondents are usually more motivated and involved. The interviews offer possibilities not offered with a self-administered questionnaire. For example, the interviewer can control the sequence of questioning by using visual displays such as concept cards, pictures, or representations of rating scales. Probing for clarification of answers or more detailed information is also possible. The major disadvantage of this method is its cost. If personal interviews are selected as the data-collection method, it is extremely important that all interviewers undergo rigorous training prior to the interviews. Dry runs should be conducted using actual consumers who are not participants in the HUT.

Self-Administered Questionnaire

In many cases, the product samples are either mailed or provided to respondents with instructions and a self administered questionnaire to be completed, at times while, and always after, the products are used. The questionnaire is mailed back to the researcher after it is completed. This method requires that the questionnaire be designed so that instructions and questions are easily understood without assistance. It is advisable in such cases for the questionnaire to have a telephone number and name of a contact person who can be reached to answer any questions. This method is less labor intensive than telephone interviews, and if conducted properly, eliminates interviewer bias; the reduced cost is an advantage.

The disadvantages of the self-administered questionnaire are that cooperation may be poorer and the response rate lower if the instructions are not well understood. The sequence of questioning cannot be controlled, as the entire questionnaire can be read prior to starting the test. Furthermore, there is no opportunity to correct errors or probe for more complete information unless the respondent telephones the contact person.

Telephone Interview

With proper planning, data may be collected by telephone interviews. Telephone interviews have the advantages of lower cost and greater speed of data collection and therefore less time, ease of maintaining random sampling procedures, maintaining geographically dispersed samples, ease of getting interviews during the important evening period, and the usually lower nonresponse rate. Compared to mail interviews, the advantages of telephone interviews are greater speed of data collection, ease of maintaining random samples, lower nonresponse rates, and ease of getting "hard-to-reach" respondents such as employed men and teens. There are several limitations of telephone interviewing. Telephone interviews, compared to personal interviews, need to be limited in time or they might be terminated; therefore, rapport with the panelist is not as good as in personal interviews, questionnaires have several constraints, such as the inability to get complete, open-ended responses as from personal interviews, and visual displays cannot be used; only rudimentary scaling can be used. Information that can be gathered through observation, such as the race of the respondents or the type of housing they live in, is not possible to obtain. It is more expensive to conduct than mailed questionnaires. The telephone interview is done to reduce costs, particularly when the respondents may be difficult to contact due to geographic constraints or availability during certain time periods.

Telephone interviews may result in poorer cooperation than in personal interviews, and the response rate may be lower than in personal interviews. Furthermore, the range of questioning is often limited, because visual displays are not possible to use, and probing, if used, may not be as effective. The interview questionnaire should be pilot-tested with a small sample of respondents, especially concentrating on questions that require the use of a scale. It is extremely important to ensure that supervisors and telephone interviewers undergo extensive training on the use of the questionnaire and the method of interviewing prior to the test. Quality-control measures that may be employed are audits of the interviewing process.

The telephone interview may originate from a central location where a bank of telephones are installed or from the interviewer's own home. The central location permits immediate supervision of interviewers to handle problems as they arise. The telephone lines may be long-distance, so that interviews may be conducted in any city desired.

Conclusion

The HUT is reserved for the later stages of sensory evaluation testing with consumers. It has a high degree of face validity, and it provides an opportunity to measure performance and acceptance by family members under normal-use conditions. It may be more costly than other sensory tests, require more time to complete, and there are no environmental controls (Stone and Sidel, 1993).

References

ASTM, Committee E-18. 1979. *ASTM Manual on Consumer Sensory Evaluation*, ASTM Special Technical Publication 682, E. E. Schaefer, ed. American Society for Testing and Materials, Philadelphia, PA. pp. 28–30.

Hashim, I. B., Resurreccion, A. V. A., and McWatters, K. H. 1995. Consumer acceptance of irradiated poultry. *Poultry Sci.* 74:1287–1294.

Hashim, I. B., Resurreccion, A. V. A., and McWatters, K. H. 1996. Consumer attitudes toward irradiated poultry. *Food Technol.* 50(3):77–80.

Meilgaard, M., Civille, G. V., and Carr, B. T. 1991. *Sensory Evaluation Techniques*, 2nd ed. CRC Press, Boca Raton, FL.

Resurreccion, A. V. A., and Heaton, E. K. 1987. Sensory and objective measures of quality of early harvested and traditionally harvested pecans. *J. Food Sci.* 52:1038–1040, 1058.

Stone, H., and Sidel, J. L. 1993. *Sensory Evaluation Practices*, 2nd ed. Academic Press, San Diego, CA.

10
Quantitative Methods by Test Location—Simulated Supermarket-Setting Tests

Introduction

Appearance attributes such as color, size, shape, and so on contribute to purchase intention. Experimentally controlled conditions for testing consumer purchase intention can be provided in simulated supermarket setting (SSS) tests. These tests allow for the testing of consumer purchase and repeat purchase behaviors under environmentally controlled conditions, including odor and lighting conditions, psychological distractions, and a comfortable testing environment.

Reasons for Conducting Simulated Supermarket-Setting Tests

Although the focus groups and sensory tests give an indication of product acceptance, it is not known whether this will lead to consumer purchase and repeat purchase. The SSS test provides a means by which purchase behavior can be measured.

Advantages and Disadvantages

Advantages

The major advantage of using a SSS test is that it allows the researcher to determine purchase behaviors under controlled conditions. If the SSS is located within the research facility, it is particularly appealing because of the accessibility of the laboratory to the researchers. In the SSS, the researchers are able to control all conditions of the test, such as product preparation and purchase, lighting, noise, and other distractions, and conduct of the test where panelists are isolated from the influence of other shoppers. The SSS test involves 50–100 consumers, the same number of panelists as in a consumer laboratory test, and fewer individuals than in a test that would be conducted in a retail store. The panelists recruited for the test would as closely as possible represent the target market for the product. In the SSS test, unlike in a retail store, it is possible to measure the effect of posters, labels, and educational intervention on the purchase of a food product.

Disadvantages

Consumer tests in the sensory laboratory are limited by the consumer panel being used. When consumers are recruited from a database, mailing lists need to be maintained. If consumers screened from a prerecruited consumer database are used, most of those who would be willing to participate are those whose homes or places of employment are situated close enough to the SSS.

Role of the Project Leader

The project leader has responsibility for facets of the test from the planning stage to the reporting of final results to the client or requester. The project leader will need to plan the design of the experiment, define the consumer panel characteristics, coordinate panelist recruitment and screening, postshopping questionnaire design and development, preparation and presentation of samples, the test procedures, testing environment, and quality control of all aspects of the test, data collection, processing, and reporting and interpretation of results.

The project leader coordinates and organizes the study. The project leader needs to understand the purpose and the goals of the study as stated by the client or requester in order to thoroughly define the problem and identify critical objectives. The project leader develops the SSS test plan to be submitted to the client or requester that delivers the information needed, outlines assumptions expected from the client, and information that can be delivered as a result of the study. During the planning stages of the SSS test, the project leader must very specifically and clearly state the requirements of the test. The project leader has major responsibility for implementing the test. There are several considerations that need to be made regarding the SSS test, including the samples and sample labels, conditions for holding the samples—whether refrigerated or at ambient temperature, setup of the supermarket, randomization of placement of samples, posters, audiovisual materials, cashiers, and other project personnel.

The project leader has several responsibilities during the planning and conduct of the SSS test. The major areas of responsibility for the project leader are to

1. Prepare a clear statement of objectives.
2. Design the test, including a brief description of the test procedure sequence and schedule.
3. Describe in detail the test procedure, food product labels, methods to be used, including preparation of samples, suggested containers, shelf conditions (refrigerated or ambient temperature), randomized placement of samples on shelves, setup of the supermarket area, posters as needed, audiovisual materials in intervention area, and any additional instructions regarding the test.
4. Identify test samples, number of test products, and number of responses per sample.
5. Design the postshopping questionnaire and pretest as needed.
6. Describe the consumer panel, panel size, and classification quotas for recruitment.
7. List screening criteria and prepare recruitment script and screener; monitor recruitment performance.
8. Specify requirements for shopping area, position of cashiers, and other special needs.
9. List food product samples and other items to be purchased.
10. Assign personnel for recruitment, greeting, preparation, and setup of samples on shelves; cashiers and for other tasks needed for the test. Identify additional personnel needs and their responsibilities.
11. Be responsible for planning and implementation of the study and quality assurance to ensure that correct testing procedures are followed.
12. Describe methods for data processing, analysis, and interpretation.
13. Decide on panelist incentives.

It is important for the project leader to monitor the performance of all personnel during all phases of testing. The project leader must ensure that the test protocol is implemented as outlined, the panelists are selected and screened as specified in the project plan, cashiers' tallies are reviewed for completeness of responses, project activities are completed according to schedule, and the quality of the test is maintained throughout the test. For example, assigned samples are placed in the supermarket shelves in the appropriate sequence at the appropriate time, and the test samples are monitored for consistency throughout the study.

The necessity for sample consistency to be maintained is especially important if testing is to be carried out over a number of days. Completed postshopping questionnaires should be examined especially at the beginning of the test so that problems can be corrected early. After this, completed questionnaires should be examined periodically. Interviews should be closely monitored to prevent interviewer bias.

Panel

For a detailed discussion on selection of a consumer panel, recruitment, and screening, please see Chapter 4. The consumer panel often consists of consumers who are recruited and screened for their eligibility to participate in the tests from a consumer database consisting of prerecruited consumers or from other sources, as described in Chapter 4. Usually, 50–100 responses are obtained during an SSS test.

Recruitment of Panelists

The SSS test requires different selection criteria for panelists compared to those in discrimination or descriptive tests. It is important in this type of test to recruit a panel that closely approximates the target market for the product.

One approach to recruitment is to recruit using a database of consumers who may be available for testing. The demographic characteristics of the individuals in the database are known and updated on a regular basis. An alternative would be to recruit respondents through market research agencies or by designated recruiters in the project team, who recruit by telephone calls from various lists of potential panelists. Other methods often used to recruit panelists are referrals, random selection from a telephone directory, random-digit dialing, posters in retail stores, mailing lists from organizations, and assorted consumer databases and intercepts at malls, shopping areas, or restaurants.

Employees versus Nonemployees

At this stage in the testing of consumer purchase behavior toward a product, company employees should not be used. This is because the food product samples are no longer tested blind, but ar often labeled and branded and, depending on the objectives of the study, may have information that would cause bias even among those employees who exhibit consumption and preference patterns similar to consumers.

Local Residents

One method of recruitment is to bring local residents to the SSS facility. This approach is growing in use, because it is convenient for the sensory staff in terms of scheduling of tests and the speed at which tests can be satisfied, and minimizes the reliance on employees. However, the method of recruiting local residents involves prerecruitment, database management, scheduling, budgeting, accessibility and security. A system must be established to contact and schedule local residents for a test. A budget and means of providing incentives or honoraria to panelists is required, which was not needed in tests involving employees. Accessibility of the SSS to panelists so that they do not wander in restricted areas is necessary. The advantages of using local residents is that participants are usually highly motivated, will show up promptly for each test, and are willing to provide considerable product information. If the SSS facility is located near or within the company premises, the disadvantage is that local residents may, for one reason or other, be biased for or against the products made by the company.

One possibility is for a company who employs this methods and other consumer testing methods that require recruitment of consumers is to develop and maintain an off-premise test facility. This would eliminate many of the problems associated with bringing local residents to the company premises. It will, however, be more costly than bringing people into an SSS facility located within company premises. Another alternative for a company is to contract with consumer and sensory evaluation agencies that have access to the necessary test resources and experience in conducting such tests.

Screening

For a given test, panelists need to qualify according to predetermined criteria that describe the target market for the product. A screener should be developed collaboratively with the sponsor of the test. The screener should be prepared, so that it ensures reliability of the screening process and provides a tally to verify quotas. Selection of participants should be conducted as rigorously as possible. The undesirable practice of using those participants who can be most conveniently contacted, such as those people who live close to the facility, should be discouraged. Similarly, the use of friends and relatives of project staff should be avoided. These individuals may bias the results of the test and should not be used. Demographic criteria such as age, gender, frequency of use of product type, availability during the test date, and other criteria such as employment with marketing firms or the sponsoring company, or similar business concerns, and other security screening criteria are often used. In addition, ethnic or cultural background, occupation, education, family income, experience, and the last date of participation in a consumer test may be used. Individuals who often participate in consumer test may no longer qualify as naive consumers. All consumers who have in-depth knowledge of the product, or those who have specific knowledge of the samples and the variables being tested, should not be included in the test.

Attendance

The problem of "no-shows" should be minimized to whatever extent possible. It reduces the panel size and is likely to present other problems, the least being a source of embarrassment with clients. To minimize the attendance problem, a number of steps can be taken. It may be necessary to overbook according to a predetermined "no-show" rate, but the rate has to be determined from attendance records from previous consumer tests. For example, recruitment averages above 20% may be high for a panel of consumers recruited from a database. Conversely, 50% or higher may be expected when the panel is prerecruited through store intercepts.

To obtain the desired participation, select participants who live approximately no more than 30 minutes from the SSS facility, and give participants clear directions and a map to help them find the location; plan test dates so as not to be in conflict with major community or school events. Send reminder letters or brightly colored postcards to post in a visible location, pay adequate fees or incentives, overbook, and, most important, make participants understand the importance of their attendance and promptness and the value of their participation in the test.

Incentives

Participants in SSS tests are usually paid for their effort. Payment is given after the session is completed. The amount of incentive paid to participants may differ according to many factors, including length of the test, location of the test and associated travel expenses, and incidence rates of qualified participants. Incentives for participation may be provided in the form of a cash honorarium, selection from a gift catalog, gift certificates, tickets to special functions such as ball games or concerts, or donation to charity or a nonprofit organization. In addition, the products purchased by the participants at the SSS are theirs to keep.

Physical Facilities

Location of the SSS Facility

The location of the SSS facility is important, because location determines how accessible it is to panelists and, as a result, influences consumer participation. The SSS should be located so that it is convenient for the majority of the test respondents. Test facilities that are not convenient to go to will not only reduce the number of consumers that will want to participate, but, more important, also limit the type of panelists that can be recruited for the tests.

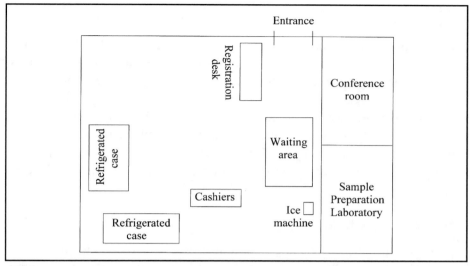

Figure 10.1 Floor plan of simulated supermarket setting (SSS) tests site. Two refrigerated cases hold the products at the desired temperature.

It is best to locate the SSS facility away from heavy traffic areas to avoid confusion and noise. For example, when the SSS facility is located within company premises, it should preferably not be situated near a noisy hallway, lobby, or cafeteria, because of the possibility of disturbance during the test. The panelists should not be able to hear the telephones ringing or telephone conversations and other office, food production or laboratory equipment. If the SSS facility is located in those areas to increase accessibility to panelists, it is preferable that the laboratory should be equipped with special soundproofing features.

Simulated Supermarket Facility

A floor plan for the SSS test site is illustrated in Figure 10.1. The SSS should be planned for efficient physical operation; furthermore, it should provide for a minimum amount of distraction to panelists from laboratory equipment and personnel, and between panelists themselves. The SSS facility should be composed of separate sample-preparation and testing areas. These areas must be adequately separated to minimize interference, during testing, due to sample-preparation operations. The sample preparation should not be visible to the panel. Sufficient supermarket shelf area is essential to avoid distraction between panelists.

Reception Room

The panel participants usually do not proceed directly to the SSS. They should be received in an area where they can hang up their coats, fill out necessary forms, and be comfortable until the orientation period. It is desirable to have a waiting area, separate from the testing and food-preparation areas where participants can register, fill out demographic, honorarium, and consent forms, and wait before or after a test, without disturbing or influencing those who are doing the test. This area allows room to encourage social interaction prior to testing, orient panelists on the test procedures, and receive payment of incentives.

Other Designated Areas

If panelists fill out postshopping questionnaires, this activity should be done in a separate area from the reception room to isolate the panelists from consumers who have not participated in the test. When the SSS test requires panelists to undergo some form of intervention, such as a slide show or video presentation, after a shopping trial, another room for this activity needs to be supplied. The conference room in Figure 10.1 may be used for the purpose.

Odor and Lighting

The SSS area should be free from any odors that would influence the results of the test. Adequate illumination is required in the SSS for reading food labels, and for examination and evaluation of food samples.

The Testing Environment

The SSS must be comfortable enough to encourage panelists to concentrate on tasks assigned. The temperature and humidity must be controlled to provide and maintain a comfortable test environment. The furnishings must be comfortable and functional. All areas should be designed for comfort and concentration during testing.

Procedure

Orientation of Panelists

The orientation of consumer panelists should consist only of describing the mechanics of the test that they need to know. Examples of such topics are orientation regarding the paperwork and the shopping procedure. If intervention procedures are to be tested, extra directions should be provided.

The orientation must be carefully planned to avoid any opportunity for altering the panelists' attitudes toward any of the food samples to be evaluated. Avoid giving any hint of the expected results of an experiment, and do not discuss the sample with the panelists prior to the test.

Test Procedure

In the SSS test, panelists are provided with a certain amount of money to purchase food products. They are asked to approach the store area and purchase food products. This usually involves purchase of one item of a product type, for example, "Purchase one package of chicken breast meat and one package of chicken thigh meat" (Hashim et al., 1995) in a study of consumer purchase of irradiated poultry, or "Purchase one package among five different poultry products" (Elsner et al., 1997). Care must be taken by the project leader to ensure that a minimum number of consumers approach the grocery store case at any one time to minimize interaction between panelists. After the panelist selects the products, these are taken to the cashiers, who accept payment and record the purchases made.

Effect of Intervention

To study the effect of educational intervention, or an advertisement, participants may then be instructed to proceed to another area for intervention, after the purchase is made, where they may view a slide show or videotape, and so forth, or sit comfortably while products are changed, and informational material such as posters or advertisements are positioned on the supermarket shelves to replace food products where gaps on the supermarket shelf have occurred for a short period. Then, the panelist is asked to make another trip to the supermarket to select another product. This process may be repeated as many times as needed, taking care to avoid causing fatigue (Elsner et al., 1997).

Ranking

It is possible to rank several products by asking the consumers to purchase only one product and pay for it. After that, panelists are asked to return to the supermarket case and purchase another product other than products they have already purchased. The consumers do not know how many trips they will take to the supermarket shelves, and they know beforehand that they will keep the products they have purchased. It is assumed that they will first purchase the product they like most, and so on, so that the last purchased product is assumed to be the product they least like.

Posttest Activities

After the SSS, the consumer maybe asked to fill out a postshopping questionnaire (Hashim et al., 1995; Elsner et al., 1997). The participants may be asked to take the product home and participate in a home-use test. After the home-use test, the participant may be invited to participate in a second SSS test to determine repeat purchase behavior (Hashim et al., 1995).

Other Considerations

Dry-Run and Briefing

It is recommended that a complete dry-run of all testing procedures be conducted on the test date, from the preparation of samples and orientation of panelists to actual test procedures to be conducted using two or three untrained individuals as panelists, one week before the test date. This will allow sufficient time for changes to be made if necessary.

References

Elsner, R. J. F., Resurreccion, A. V. A., and McWatters, K. H. 1997. Consumer acceptance of ground chicken. *J. Muscle Foods* 8:213–235.

Hashim, I. B., Resurreccion, A. V. A., and McWatters, K. H. 1995. Consumer acceptance of irradiated poultry. *Poultry Sci.* 74:1287–1294.

Hashim, I. B., Resurreccion, A. V. A., and McWatters, K. H. 1996. Consumer attitudes toward irradiated poultry. *Food Technol.* 50(3):77–80.

11

Affective Testing with Children

Physiological and Cognitive Development in Young Children

Early Preferences in Taste and Aroma

The preference for sweet foods appears early in human development. Infants have shown a liking for sweet-tasting substances through pleasant facial expressions while suckling and a dislike for salty, acidic, or bitter foods (Klaus and Klaus, 1985; Cowart, 1981; Lawless, 1985). This preference for sweeter tasting products has been shown in many studies with children (Kroll, 1990). Tuorila-Ollikainen et al. (1984) found that children put more emphasis on sweetness at the expense of other sensory attributes.

There is a difference between children and adults' odor tolerances and odor preferences (Mennella and Beauchamp, 1991). Moncrieff (1966) concluded that children ages ten, twelve, and fourteen prefer fruity smells over floral smells, whereas the opposite is true in adults. Specifically, strawberry was judged to be more pleasant by individuals under the age of twenty than individuals over twenty, whereas lavender and orange blossom were ranked more pleasant by individuals over twenty.

Cognitive Development

The understanding of children's general cognitive capabilities may provide understanding into their abilities to perform sensory tests. Children can be classified into Piaget's stages of cognitive development (Papalia and Olds, 1993; Thomas, 1992). Children ages two to seven are classified as "preoperational," which generalizes children as perception bound and limited in their logical thinking abilities. Some of the limitations on preoperational thought include the concept of "centration," which focuses on the child's ability to pay attention to one aspect of a situation at one time. Children are unable to notice the many dimensions of a situation at one time (Wadsworth, 1984). This concept is demonstrated in children's apparent concentration on one aspect of a food, such as appearance, in making their judgments, rather than all attributes of appearance, taste, color, and texture when evaluating a food product sample.

It is important to keep in mind that chronological age is only a guideline in assessing a child's cognitive abilities. The evaluation of children's ability to understand and use their senses may also be used to assess cognitive development (Maier, 1978). Using the 3-point hedonic scale, Phillips and Kolasa (1980) determined that preference for vegetables, using either pictures of food or tasting it, was essentially the same. However, lower ratings for preference were obtained when the interviewer only named the foods. These results support the contention that children require sensory stimuli to adequately rate their preferences in food.

Sensory tests often require classification activities or memory tasks. Children ages two to five years may classify items according to one characteristic, such as red color, and then may change their classification to shapes (Papalia and Olds, 1993). Fallon et al. (1984) found that children reject foods based purely on sensory characteristics—substances that taste bad are bad for you; adults reject foods based on the perception that they are dangerous, disgusting, or inappropriate. With memory tasks, studies show that children remember better if the testing situation is more game-like. Children do better at tasks if they understand the purpose or the main goal of the task so that learning can be more efficient (Brown et al., 1983). Memory capabilities may play an important role in certain sensory tests where the panelists must remember the taste sensations of three products in order to make a judgment. Studies with two-year-olds show that they can successfully perform about 80% on recognition tasks, but average less than 20% on recall tasks. Four-year-olds can do a little better; their performance is about 90% on recognition tasks, and they generally are good at recalling items that are presented last in a series., Progressively older children do better on both recognition and recall (Papalia and Olds, 1993).

Sensory Testing with Young Children

Special Problems

Sensory testing with children involves special problems not encountered with consumer panelists in older age groups. Some of these problems include (1) verbal skills (Wadsworth, 1984), (2) short attention span (Moskowitz, 1994), and (3) difficulty in comprehension of standard sensory tests by children (Moskowitz, 1985). These problems in administering sensory tests to children reflects the developmental capabilities of young children.

Verbal Skills

Children's limited verbal skills can affect understanding of the questions addressed to them. Children do not interpret and understand questions in the same manner as adults (Thomas, 1992). Studies show that children have a tendency to repeat statements by peers and adults without understanding them (Contento, 1981) because they do not have the cognitive maturity for understanding these statements. The manner in which children may answer questions depends on the phrasing of the question (Mennella and Beauchamp, 1991); thus, a positively phrased question tends to be answered in an affirmative manner.

When understanding of sensory attributes were tested, it was found that children may confuse sweetness with off-taste (Moskowitz, 1985), tartness (Moskowitz, 1985; 1994) or sourness with bitterness (Moskowitz, 1985), and also confuse attributes such as taste and texture (Moskowitz, 1994). Verbal skills begin at the age of three or four years (Thomas, 1992). Children ages five to seven are either preliterate or may have rudimentary reading skills, thus requiring personal interviews (Kroll, 1990) that are more labor intensive to conduct and incur higher costs. For these reasons, it may not be prudent to include attribute diagnostic questions when testing with young children.

Short Attention Span

Considering that many sensory tests require much concentration in order to discriminate to find a difference or indicate a preference, children's short attention spans serve as a potential problem. A lively atmosphere consisting of decorating the room with colorful decorations has been used with many studies to create a sense of fun (Kimmel et al., 1994), but tends to distract children from focusing on the sensory tasks. On the other hand, colorful decorations provided in a reception area help to relax children while waiting for their turn (Moskowitz, 1985) to participate in the test.

Difficulties in Comprehension

Due to the differences in cognitive abilities of children in the same age group, some standard sensory methods are difficult to administer unless modifications are made to accommodate children's varying cognitive abilities. When children were tested with two different types of tests (preference and difference tests) sequentially, children between the ages of two and five years experienced difficulty in adapting to new testing methodology (Kimmel et al., 1994). When children were asked to score beverages for preference and later in the test were asked to select based on sweetness intensity, many of them selected for preference in the second test rather than sweetness intensity. Most studies, however, have shown that children are able to discriminate among spices (Thomas and Murray, 1980), preferences (Kroll, 1990), sweetness (Kimmel et al., 1994), and high- and low-fat dairy products (Monneuse et al., 1991).

Training and Screening

Several methods have been used to screen and train children to be used in sensory tests. Birch and Sullivan (1991) recommend using a group demonstration on the procedure and one individual training session for children to be tested. Other methods used to determine if children understand concepts used during the tests include "same–different" tests with blocks and soft drinks (Thomas and Murray, 1980) as well as beads (Feeney et al., 1966).

Sample Size

Although the American Society for Testing and Materials (ASTM, 1992) does not have recommendations for serving sizes to be used when testing with children, they do suggest the following when testing with baby food: For the testing of baby foods, two tablespoons (approximately 60 ml) of a food product sample in a 180 ml (6 oz) cup is a sufficient amount to use. When using beverages as a test material, many different serving sizes have been used. Children have been served samples with as little as 30 ml (1 oz) in a 90 ml (3 oz) cup (Morse, 1953) and as much as 60 ml (2 oz) of a beverage (Tuorila-Ollikainen, 1984).

Age Differences in Discrimination Abilities

Sensory testing with children has been conducted using different age groups, such as five to ten (Kroll, 1990), eight to ten (Spaeth et al., 1992), two to ten (Kimmel et al., 1994), and three to four (Birch, 1979, 1981). A summary of literature on discrimination and preference methods used with children can be found in Table 11.1. As expected, children ages two to three years old are likely to understand the tasks associated with acceptance tests but not for discrimination tests, due to their inability to understand the test instructions (Kimmel et al., 1994).

Young panelists often seem confused with paired comparison and triangle tests but perform well on the duo-trio test (Feeney et al., 1966). Children often select the middle sample in a triangle test as the odd sample (Morse, 1953) or become confused with the number of samples to compare (Feeney et al., 1966). An increasing ability to perform discrimination tests is seen in older children, ages five to seven years. Children in this age group are able to successfully complete the paired comparison test for discriminatory purposes (Kimmel et

Table 11.1 Appropriateness of Sensory Testing Methods Used with Children 2 to 8 Years Old[a]

Sensory Tests	Age Groups (Yr)			
	2	**3–4**	**5–7**	**8**
Discrimination				
Paired Comparison	—	Y(2), N(9)	Y(1)	Y(1)
Triangle	N(1)	N(1,2,9)	N(2)	Y(1)
Duo-Trio	N(1)	—	Y(1), N(8)	Y(1)
Ranking for Intensity	N(1)	N(1)	Y(1)	Y(1)
Preference				
Paired Comparison	Y(1,3,4)	—	Y(2)	—
Rank Order	—	Y(5)	—	—
Hedonic scales				
3-point	—	Y(6,7,10)	—	—
5-point	—	Y(12)	—	—
7-point	—	Y(1)	—	Y(2)
9-point	—	—	Y(2)	Y(2)
Self-administered	—	—	—	Y(2)

[a]Y = Yes; N = No; 1, Kimmel et al. (1994); 2, Kroll (1990); 3, Birch (1980); 4, Johnson et al. (1991); 5, Birch (1979); 6, Birch et al. (1990); 7, Birch and Sullivan (1991); 8, Thomas and Murray (1980); 9, Feeney et al. (1966); 10, Birch et al. (1990); 11, Morse (1953); 12, Fallon et al., 1984.

al., 1994; Thomas and Murray, 1980). When the duo-trio or triangle methods were used with children in the five to seven age group, Kimmel et al. (1994) found significant results with six- and seven-year-olds, whereas Morse (1953) determined that the duo-trio was still too complicated for this age group. Children over the age of eight years were all able to perform all typical discrimination tests.

Age Differences in Performing Sensory Preference Tests

Differences in abilities to perform preference tests are seen with children of different ages. Young children, ages two to five years of age, are able to express their preference using the paired comparison method (Kimmel et al., 1994; Birch, 1981; Johnson et al., 1991).

Paired Comparison Tests

Morse (1953) used a modified paired comparison procedure to evaluate young children's preference for two types of juice. In his study, children were asked to indicate which juice they wanted more. The sample picked by the child is considered the preferred sample. Kimmel et al. (1994) found that children two years of age could reliably perform a paired-preference test under the appropriate environment and if a one-on-one verbal test protocol was used. Schmidt and Beauchamp (1988) observed that three-year-old children could reliably indicate their preference for odors using a paired test involving puppets.

Ranking

Rank ordering is another method used with young children to express preference. Birch (1979, 1980) used a modified rank-order procedure successfully with children in this age group. In this procedure, the preferred item is removed in a rank order to establish preference.

Category Scaling

The use of facial hedonic scales are popular in determining preferences. These may be simple "smiley face" scales or may depict a popular cartoon character primarily intended for use with children (Figure 11.1) and those with limited reading or comprehension skills (Lawless and Heymann, 1997). Stone and Sidel (1993) cautioned that young children may not have the cognitive skills to infer that some of face scales depicted in Figure 11.1a and b are supposed to indicate their internal responses to a test product. In studies using the facial scales, the 9-point "smiley face" scale (Figure 11.1e) did not perform well with children three to five years old (Chen et al., 1996). Although data are mixed as to the appropriate scale length to use with children in different age groups, the most popular length has been the 3-point scale (Birch and Sullivan, 1991; Phillips and Kolasa, 1980; Birch et al., 1990). Scale lengths from the 5-point scale (Fallon et al., 1984) and the 7-point scale (Kimmel et al., 1994) were used with children around four years of age. The use of unstructured scales is not recommended, because children have been observed to place their responses on the extreme ends of the scale, either "love" or "hate" a product rather than using the entire scale (Moskowitz, 1985). Chen et al. (1996) demonstrated that the 5-point facial hedonic scale with Peryam and Kroll (P&K) verbal descriptors (Kroll, 1990) could be effectively used and understood by forty-seven to fifty-nine-month- (four year-) old children, as the 7-point facial hedonic scale with P&K verbal descriptors (Kroll, 1990) could be used with sixty to seventy-one-month- (five year-) old children. The facial scales used by Chen et al. (1996) are shown in Figure 11.1c, d and e. Kroll (1990), in preference tests on children ages five to seven years old used the 9-point scale with verbal descriptors (see Table 11.2) and found this age group to use the scales effectively. In addition,

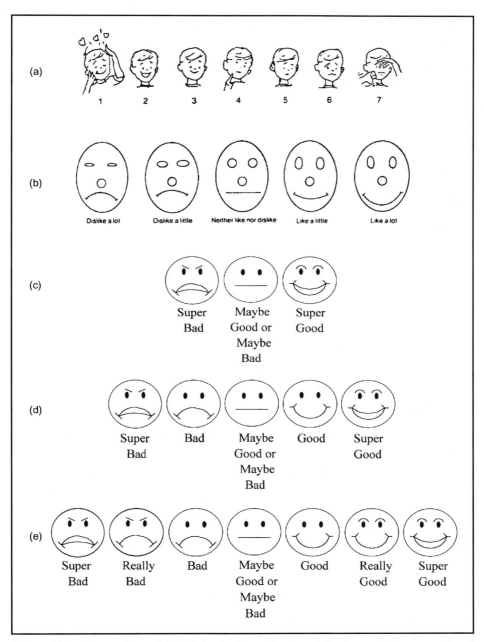

Figure 11.1 Examples of facial hedonic scales used in determining preschool children's preferences: (a) and (b), pictorial scales that are found in the literature that appear to have been used for measuring children's responses by products (reprinted from Stone and Sidel, 1993); (c), (d), (e), pictorial scale for hedonics suitable for young children (Chen et al., 1996).

Table 11.2   The P&K Verbal Scale for Affective Testing With Children (Kroll, 1990)

Super good
Really good
Good
Just a little good
Maybe good or maybe bad
Just a little bad
Bad
Really bad
Super bad

children ages eight and above are able to complete self-administered question-naires, thus eliminating the need and cost for one-on-one interviewers (Kroll, 1990).

Panel

Panel Size

The panel of children is recruited and screened for eligibility to participate in the tests, usually from a consumer database consisting of prerecruited children. Usually, 25 to 50 responses are obtained; however, as in testing with adult consumers, 50 to 100 responses are considered desirable (IFT/SED, 1981). In a test consisting of 24 panelists, it may be difficult to establish a statistically significant difference in a test with the small number of panelists. However, it is still possible to identify trends and to provide direction to the requestor. With 50 panelists, statistical significance increases to a large extent.

Recruitment of Children

Local Residents

One method of recruiting children is to bring local residents into the laboratory. This approach is growing in use because it is convenient for the sensory staff in terms of scheduling of tests and the speed at which tests can be completed satisfactorily. However, the method of recruiting children from nearby neighbor-hoods involves prerecruitment, database management, scheduling, budgeting, and accessibility. A system must be established to contact parents of potential panelists and schedule these children of local residents for the test. Methods used to recruit adult panelists may be used for children. If a database of local residents is maintained, it may contain information on the number of children in the household in each age category. Parents would be more likely to allow recruitment of their children for the test when they are familiar with testing activities of the company or sensory testing agency. A budget and means of providing incentives for the children is necessary. The advantages of using children who are local residents is that participants are usually highly motivated and will show up promptly for each test.

Another possibility is for a company to develop and maintain an off-premises test facility. This eliminates many of the problems associated with bringing local residents to the company premises. It will, however, be more costly than bringing people into the sensory laboratory, as a separate facility has to be maintained for a specific period. Another alternative is to contract the testing with a sensory evaluation research agency, assuming that it has the necessary sensory test resources.

Facilities

Testing of young children may be carried out in an interview room within the company premises or in a central location. Accessibility of the interview rooms to child panelists and their parents or caregiver, so that they do not wander in restricted areas, is necessary. If possible, each interview should be carried out in a separate room to avoid distraction from others. If this is not possible, the interviews should be conducted in a room large enough to accommodate several interviewers without distraction or interference to and from others in the same room. The interview room should be free from distractions to the child during the interview process; therefore, the use of balloons and decorations in an effort to make the child feel at ease should be confined to the reception area. In addition to the interview facilities, a waiting area should be provided for parents and caregivers. Influences from a parent or caregiver need to be minimized during the test; a room with a one-way mirror adjacent to the interview room will allow the parent to view directly the interview process. A video camera can be used to monitor the entire interview process indirectly using a monitor if direct viewing is not possible.

Furnishings

Child-sized furnishings may be used during testing with young children. Very young children (three to five years) who attend school are accustomed to working, playing, and eating on child-sized tables and chairs. Testing may be

Table 11.3 Considerations to be Made When Testing With Young Children

Use the one-on-one interview method.

Pictorial scales used with verbal scales should be used for younger children; verbal scales may be used with older children.

Scales for young children may need to be truncated.

Paired-preference tests may be conducted using very young children.

Decorations in the testing area maybe distracting to children.

conducted on these furnishings if desired. The effect of using child- or adult-sized furnishings on results of affective tests has not been determined.

Acceptance testing with children requires a few adjustments of the procedures used for testing with adults to ensure understanding and compliance, and minimize social influences (Lawless and Heymann, 1997). The considerations that should be made are as shown in Table 11.3.

References

ASTM. 1992. Standard practice for establishing conditions for laboratory sensory evaluation of foods and beverages, E480-84. American Society for Testing Materials, Philadelphia, PA, pp. 16–20.

Birch, L. L. 1979. Dimensions of preschool children's food preference. *J. Nutr. Educ.* 11(2):77–80.

Birch, L. L. 1980. Effects of peer models' food choices and eating behaviors on preschoolers' food preferences. *Child Dev.* 51:489–496.

Birch, L. L. 1981. Generalization of a modified food preference. *Child Dev.* 52:755–758.

Birch, L. L., McPhee, L., Steinberg, L., and Sullivan, S. 1990. Conditioned flavor preferences in young children. *Physiol. Behav.* 47:501–505.

Birch, L. L., and Sullivan, S. 1991. Measuring children's food preferences. *J. Sch. Health.* 61(5):212–213.

Brown, A. L., Bransford, J. D., Ferrara, R. A., and Campione, J. C. 1983. Learning, remembering, and understanding. In *Handbook of Child Psychology*, P. H. Mussen, ed. Wiley, New York, pp. 77–166.

Chen, A. W., Resurreccion, A. V. A., and Paguio, L. P. 1996. Age appropriate hedonic scales to measure food preferences of young children. *J. Sens. Stud.* 11:141–163.

Contento, I. 1981. Children's thinking about food and eating: A Piagetian-based study. *J. Nutr. Educ.* 13:S86–S90.

Cowart, B. J. 1981. Development of taste perception in humans: Sensitivity and preference throughout the life span. *Psych. Bull.* 90(1):43–73.

Fallon, A. E., Rozin, P., and Pliner, P. 1984. The child's conception of food: The development of food rejections with special reference to disgust and contamination sensitivity. *Child Dev.* 55:566–575.

Feeney, M. C., Dodds, M. L., and Lowenberg, M. E. 1966. The sense of taste of preschool children and their parents. *J. Am. Diet. Assoc.* 48:399–403.

IFT/SED. 1981. Sensory evaluation guide for testing food and beverage products. *Food Technol.* 35(11):50–59.

Johnson, S. L., McPhee, L., and Birch, L. L. 1991. Conditioned preferences: Young children prefer flavors associated with high dietary fat. *Physiol. Behav.* 50:1245–1251.

Kimmel, S. A., Sigman-Grant, M., and Guinard, J. X. 1994. Sensory testing with young children. *Food Technol.* 48(3):92–99.

Klaus, M. H., and Klaus, P. H. 1985. *The Amazing Newborn.* Addison-Wesley, Reading, MA.

Kroll, B. J. 1990. Evaluating rating scales for sensory testing with children. *Food Technol.* 44(11):78–86.

Lawless, H. 1985. Sensory development in children: Research in taste and olfaction. *J. Am. Diet. Assoc.* 85:577–583.

Lawless, H. T., and Heymann, H. 1997. *Sensory Evaluation of Food: Principles and Practices.* Chapman and Hall, New York.

Maier, H. W. 1978. *Three theories of child development*, 3rd ed. Harper and Row, New York.

Mennella, J. A., and Beauchamp, G. K. 1991. Olfactory preferences in children and adults. In *The Human Sense of Smell*, D. G. Laing, R. L. Doty, and W. Breipohl, Eds. Springer-Verlag, New York, pp. 167–180.

Moncrieff, R. W. 1966. *Odour Preferences.* John Wiley, New York.

Monneuse, M. O., Bellisle, F., and Louis-Sylvestre, J. 1991. Impact of sex and age on sensory evaluation of sugar and fat in dairy products. *Physiol. Behav.* 50:1111–1117.

Morse, R. L. D. 1953. Exploratory studies of preschool children's taste discrimination and preference for selected citrus juices. *Proceedings Florida State Horticultural Society*, November 3–5, 66:292–301.

Moskowitz, H. 1985. *New Directions for Product Testing and Sensory Analysis of Food.* Food and Nutrition Press, Westport, CT.

Moskowitz, H. 1994. *Food Concepts and Products: Just-in-Time Development.* Food and Nutrition Press, Westport, CT.

Papalia, D. E., and Olds, S. W. 1993. *A Child's World: Infancy through Adolescence*, 6th ed. McGraw-Hill, New York.

Phillips, B. K., and Kolasa, K. K. 1980. Vegetable preferences of preschoolers in day care. *J. Nutr. Educ.* 12(4):192–195.

Schmidt, H. J., and Beauchamp, G. K. 1988. Adult-like odor preference and aversions in three-year-old children. *Child. Dev.* 59:1136–1143.

Spaeth, E. E., Chambers, E., and Schwenke, J. R. 1992. A comparison of acceptability scaling methods for use with children. In *Product Testing with Consumers for Research Guidance: Special Consumer Groups*. L. S. Wu and A. E. Gelinas, Eds. American Society for Testing Materials (ASTM), Philadelphia, PA, pp. 65–77.

Stone, H., and Sidel, J. L. 1993. *Sensory Evaluation Practices*, 2nd ed. Academic Press, New York.

Thomas, R. M. 1992. *Comparing Theories of Child Development*, 3rd ed. Wadsworth, Belmont, CA.

Thomas, M. A., and Murray, F. S. 1980. Taste perception in young children. *Food Technol.* 34:38–41.

Tuorila-Ollikainen, H., Mahlamaki-Dultanen, S., and Kurkela, R. 1984. Relative importance of color, fruity flavor, and sweetness in the overall liking of soft drinks. *J. Food Sci.* 49:1598–1600.

Wadsworth, B. J. 1984. *Piaget's Theory of Cognitive and Affective Development*, 3rd ed. Longman, New York.

12

Statistical Analysis Methods

Introduction

Consumer sensory evaluation uses statistics to determine whether responses from a group of consumers are sufficiently similar or represent a random occurrence. Knowledge that results are not a random occurrence enables the project leader to make a decision about the products being tested with some measure of confidence.

The author assumes that the reader is familiar with statistical procedures commonly used in sensory evaluation. For those who wish to review the procedures, statistical methods in sensory evaluation is the topic of several reference books (Gacula and Singh, 1984; O'Mahony, 1982; Meilgaard et al., 1988). Other references on the various statistical methods are those by Cochran and Cox (1957), Petersen (1986), and Snedecor and Cochran (1980).

Hypothesis Testing

Hypothesis testing is "an approach for drawing conclusions about a population, as a whole, based on the information contained in a sample of items from that population" (ASTM, 1996). Hypothesis testing involves the development of a null hypothesis and an alternative hypothesis. The null hypothesis states the conditions that are assumed to exist before the study is run (ASTM, 1996). In comparing means of two samples, the null hypothesis is

$$H_0: \mu_1 = \mu_2.$$

This means that there is no difference between the samples on the average. The alternative hypothesis states the conditions that are of interest to the investigator if the null hypothesis is not true (ASTM, 1996). In comparing two samples, 1 and 2, the alternative hypothesis is that

$$H_a: \mu_1 \neq \mu_2.$$

Statistical Error

In data analysis of consumer affective responses, the question most frequently asked is whether a difference exists between means of responses to products. In sensory evaluation, it is an accepted practice to consider a difference as statistically significant when the value for α is less than or equal to 0.05. In accepting this difference, we are confident that there is only one chance in twenty that this result was due to chance. Similarly, if we are unable to accept this level of risk, we may set the value of α at .01, which gives us the confidence that there is only one chance in 100 that this result is due to chance.

Taking risk into account acknowledges the possibility that an incorrect decision exists. If we reject the null hypothesis when it is true, such as when we conclude that there is a difference when this is not true, this constitutes a Type I error. The probability of making a Type I error is α (the significance level). Typical values for α are .10, .05, and .01.

When the null hypothesis is true, the probability of making the correct decision is $1-\alpha$ (ASTM, 1996). If, on the other hand, we accept the null hypothesis when it is false, a Type II error occurs. In failing to reject the null hypothesis when it is false, the probability of this error is β. A mathematical relationship occurs between α and β such that when one increases, the other decreases, but the relationship is determined by the number of judgments and the magnitude of difference between products (Stone and Sidel, 1993).

Statistical Power

The statistical power of the test should also be taken into account (Stone and Sidel, 1993). The power of a hypothesis test is the probability of detecting a difference of a specified size where: If α is set very low, statistical power will be small and β risk will be high. Power = $100 (1 - \beta)\%$ (ASTM, 1996).

The number of judgments or consumers participating in the consumer test and the magnitude or degree of difference between the products influence risk and statistical power. The larger the number of participants in the consumer test, the higher is the likelihood that a difference in the samples will be detected and the β risk will be minimized. However, together with the decreasing statistical risk is the problem of "obtaining differences that have no practical value" (Stone and Sidel, 1993).

The difference in products is a source of differences independent of the variable being tested. Products are often a greater source of variability than are consumers. The use of replication is helpful in obtaining a measure of product variability. Therefore, by using more product, responses represent the range of variability within the product being tested. In addition, other procedures or controls that can be applied to eliminate nontest variables will increase power. This means strict compliance with proper sample preparation and serving methodology, use of an appropriate test site with environmental controls, and so on (Stone and Sidel, 1993).

Selection of Computer Software

The use of computer software for statistical analysis of sensory evaluation data is widespread. Many statistical software packages are available that include most of the statistical methods used in analyzing consumer test results. Not all statistical evaluation computer software is easy to use. Care must be exercised in the selection of a statistical software for data analysis.

It is usually best to read the existing consumer sensory evaluation literature or ask an experienced colleague for a recommendation regarding statistical software packages. Some software packages have the simplest to the most complex statistical capability. Some software can handle only a small number of statistical methods, and others have a larger selection to offer. While some packages result in pages of computer output and graphics, others may result in summaries that are meaningless to some users. For example, some of the software packages may have only the univariate procedures, while others have included in the package or as an option a wide selection of multivariate analysis methods that may or may not be useful in analyses.

Graphics capability is important. As much as possible, the software should have the capability of producing graphs of publication quality. If such graphics capability is not available, the software package should produce results that have the capability of being imported to a graphics package.

Statistical Analysis of Consumer Data

Exploratory Data Analysis

Data from consumer affective tests are often difficult to understand. Graphical representations of the data are often useful in increasing comprehension. Exploratory data analysis should always be conducted to provide an initial evaluation of the data. The primary purpose of exploratory data analysis is to obtain a quick look at the data to determine whether they should be analyzed further and to determine potential directions and methods for final statistical analysis. The analysis methods may be numerical and graphical (Jones, 1997). Examples of the analysis include many familiar methods such as bar charts, histograms, and scatter plots. Frequency analysis and measures of central tendency, such as the means and standard deviations, medians, and modes, provide for an assessment of any skewed response patterns and order effects, and are likewise helpful in providing the basis for decisions regarding the quality of the data. Exploratory data analysis helps in looking at distributions of the data that differ greatly from the normal distribution. Data transformations (such as taking the square roots or logarithms) are likewise a part of exploratory data analysis.

The mean or average is calculated. Because it is not usually possible to obtain the mean of a population, the best estimate of a population mean is X. Thus, the mean is the measure from which inferences are drawn about the population. The median is the middle number in a set of numbers arranged in order, such as smallest to largest. The mode is the most frequently occurring value. A measure of the spread of numbers is also useful in understanding data. The range is the absolute difference between the largest and the smallest numbers in the sample.

Graphic Representations of the Data

In many cases, it is advisable as a first step to plot the data. Graphing independent and dependent variables is a simple and direct way to visualize the nature of the relationship between variables. Scatter plots, bar graphs, and histograms are especially helpful in this task. Graphs show whether a relationship exists, or whether the relationship is a linear or curvilinear one. Outliers can likewise be detected by graphical representations of the data.

t-test

The Student's t-test is one of the most commonly used statistical procedures for determining the significance of the difference between means of two samples. The t-statistic is the ratio of the difference to the standard error of that difference (ASTM, 1996). It is a useful test when only two products are being tested and when analyzing responses from a small number ($N \leq 30$) of consumers. The use of the T-test has declined in popularity with the increased need for multiproduct comparisons (Sidel and Stone, 1993).

One characteristic of the t-statistic is that it provides information on the direction of the difference. This information is often important in the interpretation of results (ASTM, 1996). Tables of the t-distribution give the appropriate t-statistic for given probabilities and degrees of freedom. There are several variations of the t-test (O'Mahony, 1986). These are two-sample tests with related samples, two-sample tests with unrelated samples, and one-sample tests.

Two-Sample Tests with Related Samples

This is also called the paired or dependent t-test and tests for a significant difference between the means of two related samples. This test is appropriate in tests involving two samples when the same consumer panelists evaluate both samples. When the effect of the serving order of samples is important, the order of presentation of both samples is balanced such that the number of times one sample is presented as the first sample is equal to the number of times it is presented as the second sample.

Two-Sample Tests with Independent Samples

This is also called the unpaired, independent t-test (O'Mahony, 1986) or the generalized t-test (ASTM, 1996). This test is for a significant difference between means of two unrelated samples, such as those responses obtained from different judges under different conditions, and results in two independent groups of data. The data sets may or may not have an equal number of observations in each group, but they are assumed to have normal distributions and the same variances.

One-Sample Test

This tests whether a sample with a given mean came from a population with a known mean. The question asked is "Is the sample mean different from the population mean?" In some cases, this test is used to compare the average set of results against some fixed value, such as a target or specification. The calculations are similar to those for the generalized t-test.

The selection of the appropriate t-statistic and degrees of freedom will depend on a number of factors, including paired or unpaired varieties, equal or unequal numbers of judgments per cell, and equality or inequality of variances. Each alternative entails some computational differences; therefore, it is necessary that the project leader be familiar with them (Stone and Sidel, 1993). Detailed examples of the different variations of the t-test are presented by O'Mahony (1986).

Chi-Square Test

This is a method to test hypotheses about frequency of occurrence (O'Mahony, 1986) or to determine whether the distribution of observed frequencies of a categorical variable (either nominal or ordinal) differs significantly from the distribution of frequencies that are expected according to some hypothesis. The chi-square test is a nonparametric test, because it uses nominal data. Chi-square analysis is discussed in greater detail in Chapter 2 on sensory test methods.

Analysis of Variance

The analysis of variance, often referred to as ANOVA or AOV, is probably the most frequently used method for data analysis of consumer sensory data from multiproduct tests. The method is used to test for significant differences in means of a variable across groups of observations. While "analysis of means" appears to be a more appropriate name, the methods employ ratios of variances to determine whether the means differ—thus, the name analysis of variance.

The analysis depends upon the experimental design and can be simple or complex. For complex analyses, consultation with a statistician is recommended (ASTM, 1996). The computation for analysis of variance of consumer test data is usually performed using a computer because of the large data sets resulting from consumer tests.

The total amount of variation in a test can be split into different sources of variability, such as product-to-product variation, subject-to-subject variation, and within-subject variation. Some of these components represent planned differences and are called fixed effects (treatments, factors), and others are random effects such as measurement error (ASTM, 1996). ANOVA is a statistical procedure designed to partition all the sources of variability in a test, thus providing a more precise estimate of the variable being studied (Stone and Sidel, 1993). If the variance among fixed effects exceeds the variation within such effects, the fixed effects are said to be statistically different.

There are a number of ANOVA procedures that can be used. The selection of the procedure depends on the nature of the problem. The specific test procedure to use depends on whether the variable is expected to have more than a single effect, whether the subjects might be expected to respond differently to the different products, whether subjects evaluate each product on more than a single occasion, or whether different subjects evaluate the products at different times. For each experimental design, the particular ANOVA model will have certain distinguishing features. In most instances, these features are associated with partitioning the sources of variability and in determining appropriate error terms to derive the F value (the ratio of variances) for estimating statistical significance (Stone and Sidel, 1993).

The t-test is generally not thought of as an analysis of variance. However, it is a special case of one of the simplest analysis of variance procedures. The t-value is a ratio of two variances, and this ratio gives another way of telling whether there is a statistically significant difference between groups (O'Mahony, 1986; Iversen and Norpoth, 1987).

ANOVA involves a series of computations to yield total sums of squares, treatment sums of squares, and error sums of squares of the experimental observations. When the data set is not balanced, as when there are unequal numbers of subjects and therefore missing observations, the missing observations need to be accounted for. In such cases, the general linear model (GLM) procedure may be used.

Replication

When replications are used, decisions will need to be made before any computations are started regarding the most useful way to partition or group the treatment and the error sums of squares (Stone and Sidel, 1993). Replication is extremely important to provide a measure of consumer panelist reliability and to increase the statistical power of the test or, in other words, "increase the likelihood of finding a difference" (Stone and Sidel, 1993). Replications will increase the likelihood of finding a statistical difference. The expectation is that consumer variability within will be smaller than the variability across consumers.

The next step is to compare the various sums of squares after dividing each value by its associated degrees of freedom. If the responses (to all the products) are from the same population, the ratio will be equal to 1. If the between-treatments variance is large compared to the within-treatments variance (the ratio is > 1), the more likely we are to consider the responses as due to a true treatment effect. The mean square for the between-samples estimate divided by the mean square for the within-samples (or error), referred to as the F ratio, is then compared with tabulated values (Appendix E.3) for determining statistical significance.

ANOVA Table

It is common practice to summarize the computations listing the sources of variation, degrees of freedom, sums of squares, mean squares, F values and associated probability values. Examples of such computations can be found in Gacula and Singh (1984) and O'Mahony (1986).

User-friendly software to perform ANOVA is readily available for use on both personal and mainframe computers. This increasing reliance on available statistical packages needs to be treated with caution. The project leader for a consumer study must be familiar with the various ANOVA designs and how to read and interpret results. The proper partitioning of the sources of variability is critical in identifying and isolating sources such as interaction that could mask a true treatment effect. For example, using software packages where only one type of ANOVA is available (fixed effects) may lead to erroneous conclusions about differences if the particular design was a mixed design involving fixed and random effects. Split-plot and other complex designs require even more specific information to ensure that the computations and the test for significance will be appropriate. It is advisable for the project leader to consult with a statistician when more complex designs are used. The author suggests reading O'Mahony (1986), the basis for most of the discussion on ANOVA in this chapter.

Table 12.1 Analysis of Variance Table for a One-Factor Completely Randomized Design Consumer Test[a]

EXPERIMENTAL DESIGN

Treatment (*b* = 3)	Fried (A)	Baked (B)	Hot Air (C)
Panelist scores in treatment (*n* = 50)	X1	X51	X101
	X2	X52	X102
	X3	X53	X103
	X4	X54	X104
	X5	X55	X105
	.	.	.
	.	.	.
	X50	X100	X150
Grand Total = $T = T_A + T_B + T_C$	T_A	T_B	T_C
$N = b \times n = 150$			

ANOVA TABLE

Source	Sum of Squares (*SS*)	Degrees of Freedom (*df*)	Mean Square (*MS*)	F-ratio
Total	$SS_T = \Sigma X^2 - C$	$(N - 1)$		
Between Treatments	$SS_B = [T_A^2 + T_B^2 + T_C^2/n] - C$	$(b - 1)$	$MS_B = SS_B/(b - 1)$	MS_B/MS_E
Error or Residual	$SS_E = SS_T - SS_B$	$(N - 1) - (b - 1)$	$MS_E = SSE_E / (N - 1) - (b - 1)$	

[a]Where treatment = *b*, panelist scores in treatment = *n*, individual scores = X, and the total number of observations, $N = b \times n$.

Mean Comparison Tests

ANOVA provides evidence that a significant difference exists, but does not give an indication of how the treatments differ. To determine which treatments are significant, a mean comparison test is needed. Several tests are used for this purpose. These are the Fisher's LSD (least significant difference), Duncan's multiple range, and the Newman-Keuls, Tukey, Scheffe, and Bonferroni tests. It is important to remember that these tests are not interchangeable and apply only when a significant *F* value was found. O'Mahony (1986) wrote an excellent discussion on multiple comparisons.

One-Factor Completely Randomized Design

The one-factor completely randomized design (Table 12.1) involves the scoring of each treatment by different panelists. For example, in a study with three treatments, such as a new snack chip product processed three ways, Treatment A is fried, Treatment B is baked, and Treatment C was cooked in hot air. Treatment A was evaluated by 50 panelists, Treatment B by another 50 panelists, and Treatment C by yet another 50 panelists, for a total of 150 panelists. ANOVA is used to determine whether the treatments have an effect on the consumer panelists' scores or whether the treatment means are different. The null hypothesis states that the treatment means are equal. The ANOVA table for this test is as shown in Table 12.1 where

The correction term (O'Mahony, 1986) is $C = T^2/N$,
   where $T = T_A + T_B + T_C$
The total sum of squares $SS_T = \Sigma X^2 - C$, with degrees of freedom,
   $df = N - 1$;
The between-treatments sum of squares,

$$SS_B = \frac{T_A^2 + T_B^2 + T_C^2}{n} - C, \text{ with degrees of freedom, } df = b - 1;$$

The error sum of squares,

$$SS_E = SS_T - SS_B$$

The error term, or the difference between an observation and the mean for each group, is calculated to represent the effect of all other variables (Iversen and Norpoth, 1987). The error term is sometimes referred to as the residual term, and the error mean square is sometimes called the residual mean square. Residual means the variance left over once the variance due to the treatments has been accounted for. It is another way of thinking of the experimental error, or the variation in scores not due to any manipulation of the conditions of the experience (O'Mahony 1986).

When the number of panelists is different for each treatment, as shown in Table 12.2, a slight difference in computation is required. Although the computation of the F ratio is similar, the computation of the between-treatments sum of squares is slightly different from

$$SS_B = \frac{T_A^2 + T_B^2 + T_C^2}{n} - C$$

which is used when the number of panelists evaluating each treatment are of equal size. When the number of panelists evaluating each treatment varies, the between-treatment sum of squares is calculated by dividing each treatment total by the number of panelists evaluating that sample,

Table 12.2 Analysis of Variance Table for a One-Factor Completely Randomized Design Consumer Test[a]

	EXPERIMENTAL DESIGN		
Treatment ($b = 3$)	**Fried (A)**	**Baked (B)**	**Hot Air (C)**
Panelist scores in treatment (n)	X1	X46	X96
	X2	X47	X97
	X3	X48	X98
	X4	X49	X99
	X5	X50	X100
	.	.	.
	.	.	.
	.	.	.
	X45	X95	X150
	$n_A = 45$	$n_B = 50$	$n_C = 55$
Grand Total $= T = T_A + T_B + T_C$	T_A	T_B	T_C

[a] Where treatment $= b$, panelist scores in treatment $= n$, individual scores $=$ X, the total number of observations, $N = n_A + n_B + n_C$, and the n for each treatment is unequal.

$$SS_B = \left[\frac{T_A^2}{n_A} + \frac{T_B^2}{n_B} + \frac{T_C^2}{n_C} \right] - C$$

Computation of all other terms remains unchanged as follows:

The correction term (O'Mahony, 1986) remains $C = T^2/N$,
Total sum of squares $SS_T = \Sigma X^2 - C$, with degrees of freedom, $df = N - 1$,
Between-treatments df is as before, $df = b - 1$,
Error sum of squares, $SS_E = SS_T - SS_B$.

Computation for the df_E, was formerly

$$df_E = (N - 1) - (b - 1) = N - b,$$

however, when the number of panelists differ, df_E is calculated as:

$$df_E = (n_A - 1) + (n_B - 1) + (n_C - 1).$$

Multiple-Factor Designs and Analyses

Unlike the one-factor completely randomized design, the treatments by panelists (Trt × Panelist) design is more frequently used in sensory evaluation where each of the treatments are evaluated once by each consumer panelist. For example, if the object is to determine whether a significant difference exists between three products using 100 consumer panelists, the products will be served to each consumer in a balanced, sequential monadic order until all three products are served and evaluated. In the one-factor completely randomized design discussed previously, each panelist evaluates only one product; therefore, to obtain 100 responses for each of three products, 300 panelists would be required. The treatment by panelists design is preferred because it (1) not only provides a more effective use of subjects, but also more importantly, (2) it results in a separate panelist term that can be partitioned from the error. The treatments by panelist design can be expected to result in an increased likelihood for finding a difference. An example of the treatments by panelist design without interaction is shown in Table 12.3. There are b treatments corresponding to A, B, and C; therefore, $b = 3$, and n scores (X values) in each treatment. The n scores come from n panelists (P_1, P_2, P_3, . . . , P_n) who are tested under each treatment. There are $N = nb$ scores in the test. The totals for scores under treatments A, B, and C are the column totals T_A, T_B, and T_C. The total for each panelist are the row totals T_1 for P_1, T_2 for P_2, T_3 for P_3, and T_n for P_n. The grand total is T.

An ANOVA table for the treatments by panelist design is shown in (Table 12.3), wherein:

The correction term is $C = T^2/N$,
Total sum of squares $SS_T = \Sigma X^2 - C$ with degrees of freedom, $df = N - 1$,
Between-treatments sum of squares $SS_B = [(T_A^2 + T_B^2 + T_C^2)/n] - C$;
   $df = b - 1$.
A between-panelists sum of squares, SS_p can be calculated as follows:

$$SS_P = \frac{T_1^2 + T_2^2 + T_3^2 + T_4^2 \ldots + T_n^2}{b} - C, \text{ with } df = n - 1.$$

The denominator is b rather than n.
The error sum of squares,

$$SS_E = SS_T - SS_B - SS_P, \text{ with } df_E = (N - 1) - (b - 1) - (n - 1)$$

When a significant difference is found, multiple comparison tests can be used.

Two-factor ANOVA with Interaction

The two-factor design without interaction leaves the effects of interaction in the error term. The two-factor design with interaction partitions out the interac-

Table 12.3   Analysis of Variance Table for a Multiple Factor without Interaction Consumer Test[a]

EXPERIMENTAL DESIGN

Treatment ($b = 3$)	Fried (A)	Baked (B)	Hot Air (C)	Total
Panelist scores in treatment (n)				
P1	X1	X1	X1	T1
P2	X2	X2	X2	T2
P3	X3	X3	X3	T3
P4	X4	X4	X4	T4
P5	X5	X5	X5	T5
.	.	.	.	.
.	.	.	.	.
.	.	.	.	.
P100	X100	X100	X100	T100
Grand Total $= T = T_A + T_B + T_C$	T_A	T_B	T_C	
Mean	X_A	X_B	X_C	
$N = bn = 300$				

ANOVA TABLE

Source	Sum of Squares (SS)	Degrees of Freedom (df)	Mean Square (MS)	F-ratio
Total	$SS_T = \sum X^2 - C$	$(N - 1)$		
Between Treatments	$SS_B = [T_A^2 + T_B^2 + T_C^2/n] - C$	$(b - 1)$	$MS_B = SS_B/(b - 1)$	MS_B/MS_E
Between Panelists	$SS_P = [(T_1^2 + T_2^2 + T_3^2 + T_4^2 + \ldots + T_n^2/b] - C$	$(n - 1)$	$MS_P = SS_P/(n - 1)$	MS_P/MS_E
Error or Residual	$SS_E = SS_T - SS_B - SS_p,$	$(N - 1) - (b - 1) - (n - 1)$	$MS_E = SS_E/(N - 1) - (b - 1) - (n - 1)$	

[a] Where treatment $= b$, panelist scores in treatment $= n$, individual scores $= X$, and the total number of observations, $N = b \times n$.

tion from the error term so that it can be examined as a separate entity. The two-factor ANOVA without interaction means that the interaction is not calculated, not that there is no interaction; conversely, the two-factor ANOVA with interaction means that the interaction is calculated, not that interaction occurs (O'Mahony, 1986).

Products A, B, and C in the previous example, were tested in two cities, a and b. An ANOVA table for the treatments by panelists design with interaction is shown in Table 12.4, wherein:

Table 12.4 Analysis of Variance Table for a Multiple Factor with Interaction Consumer Test[a]

EXPERIMENTAL DESIGN

Treatment (b = 3)	Fried (A)	Baked (B)	Hot Air (C)	Total
City a:				
Panelists				
P1	X1	X1	X1	T1
P2	X2	X2	X2	T2
P3	X3	X3	X3	T3
P4	X4	X4	X4	T4
P5	X5	X5	X5	T5
.	.	.	.	.
P50	X50	X50	X50	T50
City b:				
P51	X51	X51	X51	T51
P52	X52	X52	X52	T52
P53	X53	X53	X53	T53
P54	X54	X54	X54	T54
P55	X55	X55	X55	T55
.	.	.	.	.
P100	X100	X100	X100	T100
Grand Total = $T = T_A + T_B + T_C$	T_A	T_B	T_C	T
Mean	X_A	X_B	X_C	
$N = nb = 300$				

ANOVA TABLE

Source	Sum of Squares (SS)	Degree of Freedom (df)	Mean Square (MS)	F-ratio
Total	$SS_T = \Sigma X^2 - C$	$(N - 1)$		
Between Treatments (ABC)	$SS_A = [T_A^2 + T_B^2 + T_C^2/n_A] - C$	$(b - 1)$	$MS_A = SS_A/(b - 1)$	MS_A/MS_E
Between Cities (ab)	$SS_a = [(T_a^2 + T_b^2/n_A] - C$	$(b_a - 1)$	$MS_a = SS_a/(b_a - 1)$	MS_a/MS_E
Interaction (ABC) × (ab)	$SS_{a \times A}$ = Cell total SS − SS_A − SS_a	$df_{a \times A}$	$MS_{a \times A} = SS_{a \times A}/df_{a \times A}$	$MS_{a \times A}/MS_e$
Error or Residual	$SS_E = SS_T - SS_A - SS_a - SS_{a \times A}$	$df_E = (N - 1) - (b_A - 1) - (b_a - 1) - (b_A - 1)(b_a - 1)$	$MS_E = SS_E/(N - 1) - (b_A - 1) - (b_a - 1) - (b_A - 1)(b_a - 1)$	

[a]Where treatment = b, panelist scores in treatment = n, individual scores = X, and the total number of observations, $N = b \times n$.

The correction term (O'Mahony, 1986) is $C = T^2/N$,

Total sum of squares $SS_T = \Sigma X^2 - C$, with degrees of freedom, $df_T = N - 1$,

where $N =$ the total number of scores. The between-treatments sum of squares SS_A, for treatments ABC:

$$SS_A = \frac{T_A^2 + T_B^2 + T_C^2}{n_A} - C,$$

where n_A is the number of scores from the original data set added together to give the totals T_A, T_B, and T_C; $df_A = b_A - 1$, where $b_A =$ the number of treatments, ABC.

The sum of squares, SS_a, for cities a and b:

$$SS_a = \frac{T_a^2 + T_b^2}{n_a} - C$$

where n_a is the number of scores added together to give the totals T_a and T_b; and $df_a = b_a - 1$, where $b_a =$ the number of treatments, ab.

The total cell sum of squares is:

$SS = SS_A + SS_a + SS_{a\times A}$, where $SS_{a\times A}$ is the interaction sum of squares.

To obtain the interaction sum of squares, $SS_{a\times A,}$ SS can be calculated from the matrix and, SS_A and SS_a subtracted.

The cell SS: Cell $SS = \dfrac{(T_{aA}^2 + T_{aB}^2 + T_{aC}^2 + T_{bA}^2 + T_{bB}^2 + T_{bC}^2)}{n_{\text{cell}}} - C$, and cell $df_a =$ the number of cells in the total data set $- 1$.

The interaction sum of squares is then obtained by subtraction:

$SS_{a\times A} =$ Cell total $SS - SS_A - SS_a$; and the interaction df, $df_{a\times A}$, is likewise obtained by subtraction, $df_{a\times A} = df_{\text{cell}} - df_A - df_a$ or by merely multiplying df for the factors involved in the interaction, or $df_{a\times A} = df_a \times df_A$.

Error is obtained by subtracting all terms from the total,

$$\text{Error } SS_E = SS_T - SS_A - SS_a - SS_{a\times A};$$

and $df_E = df_T - df_A - df_a - df_{a\times A}$.

Replications

The error can be improved with the addition of replication. If the interaction is nonsignificant, it need not be pooled with residual error for testing treatment effects.

Table 12.5 Analysis of Variance Table for a Three-Factor with Interaction Consumer Test

	EXPERIMENTAL DESIGN						
Treatment (b = 3)	**Fried (A)**		**Baked (B)**		**Hot Air (C)**		
Panelists	**Gas flushed**	**Control (no gas flush)**	**Gas flushed**	**Control (no gas flush)**	**Gas flushed**	**Control (no gas flush)**	**Total**
City a:							
P1	X1	X1	X1	X1	X1	X1	T1
P2	X2	X2	X2	X2	X2	X2	T2
P3	X3	X3	X3	X3	X3	X3	T3
P50	X50	X50	X50	X50	X50	X50	T50
	.	.	.	.	.	.	.
City b:							
P51	X51	X51	X51	X51	X51	X51	T51
P52	X52	X52	X52	X52	X52	X52	T52
P53	X53	X53	X53	X53	X53	X53	T53
P100	X100	X100	X100	X100	X100	X100	T100
	.	.	.	.	.	.	.
City c:							
P101	X101	X101	X101	X101	X101	X101	T101
P102	X102	X102	X102	X102	X102	X102	T102
P103	X103	X103	X103	X103	X103	X103	T103
P150	X150	X150	X150	X150	X150	X150	T150
Total $N = nb = 600$							

Testing for treatment effects may be achieved by using the residual, or in the situation of a nonsignificant interaction, by pooling the residual with the interaction sum of squares and the degrees of freedom (*df*) and using the resulting error term.

Three-Factor ANOVA Model

The next example (Table 12.5) shows the use of a three-factor ANOVA model. Products A, B and C were packaged with or without a gas flush, α and β, and tested in cities a, b and c. All 150 panelists evaluated each of the six samples. The use of six samples is for illustration only and is not recommended for consumers in one session. An ANOVA table for the treatments by panelist design with interaction is shown below wherein:

The correction term (O'Mahony, 1986) is $C = T^2/N$,

Total sum of squares $SS_T = \Sigma X^2 - C$, with degrees of freedom,

$df_T = N - 1$, where N = the total number of ratings.

Table 12.5 *Continued*

ANOVA TABLE

Source	Sum of Squares (SS)	Degrees of Freedom (df)	Mean Square (MS)	F-ratio
Total	$SS_T = \sum X^2 - C$	$(N - 1)$		
Between Treatment 1 (ABC)	$SS_A = [(T_A^2 + T_B^2 + T_C^2/n_A] - C$	$(b_A - 1)$	$MS_A = SS_A/(b - 1)$	MS_A/MS_E
Between Treatment 2 ($\alpha\beta$)	SS_α (for treatments α and β) = $[(T_\alpha^2 + T_\beta^2/n_\alpha] - C$	$(b_\alpha - 1)$	$MS_\alpha = SS_\alpha/(b - 1)$	MS_α/MS_E
Between Cities (abc)	$SS_a = [(T_a^2 + T_b^2 + T_C^2)/n_a] - C$	$(n_a - 1)$	$MS_a = SS_a/(n_a - 1)$	MS_a/MS_E
Interaction (ABC) × ($\alpha\beta$)	$SS_{A\times\alpha} =$ Cell total $SS - SS_A - SS_\alpha$	$df_{A\times\alpha}$	$MS_{A\times\alpha} = SS_{A\times\alpha}/df_{A\times\alpha}$	$MS_{A\times\alpha}/MS_E$
Interaction (ABC) × (abc)	$SS_{A\times a} =$ Cell total $SS - SS_A - SS_a$	$df_{A\times a}$	$MS_{A\times a} = SS_{A\times a}/df_{A\times a}$	$MS_{A\times a}/MS_E$
Interaction ($\alpha\beta$) × (abc)	$SS_{\alpha\times a} =$ Cell total $SS - SS_\alpha - SS_a$	$df_{\alpha\times a}$	$MS_{\alpha\times a} = SS_{\alpha\times a}/df_{\alpha\times a}$	$MS_{\alpha\times a}/MS_E$
Interaction (ABC) × ($\alpha\beta$) × (abc)	$SS_{A\times\alpha\times a} =$ Cell total $SS - SS_A - SS_\alpha - SS_a - SS_{A\times\alpha} - SS_{A\times a} - SS_{\alpha\times a}$	$df_{A\times\alpha\times a}$	$MS_{A\times\alpha\times a} = SS_{A\times\alpha\times a}/df_{A\times\alpha\times a}$	$MS_{A\times\alpha\times a}/MS_E$
Error or Residual	$SS_E =$ Cell total $SS - SS_A - SS_a - SS_\alpha - SS_{A\times\alpha} - SS_{A\times a} - SS_{\alpha\times a} - SS_{A\times\alpha\times a}$	$df_E =$ total cell $df\ df_A - df_a - df_\alpha - df_{A\times\alpha} - df_{A\times a} - df_{\alpha\times a} - df_{A\times\alpha\times a}$	$MS_E = SS_E/df_E$	

The between-treatments sum of squares SS_A (for treatments ABC),

$$SS_A = \frac{T_A^2 + T_B^2 + T_C^2}{n_A} - C,$$

where n_A is the number of scores from the original data set added together to give the totals T_A, T_B and T_C;

$$df_A = b_A - 1, \text{ where } b_A = \text{ the number of treatments, } ABC$$

$$SS_a \text{ (for treatments } a, b \text{ and } c) = \frac{T_a^2 + T_b^2 + T_c^2}{n_a} - C,$$

where n_a is the number of scores added together to give the totals T_a (for treatments a and b) + T_b, $df_a = b_a - 1$;

$$SS_\alpha \text{ (for treatments } \alpha \text{ and } \beta) = \frac{T_\alpha^2 + T_\beta^2}{n_\alpha} - C,$$

where n_α is the number of scores added together to give the totals T_α and T_β; and $df_\alpha = b_\alpha - 1$, where b_α = the number of treatments, $\alpha\beta$.

Two-Way Interactions. To compute the interaction sum of squares, between ABC and abc, $SS_{A \times a}$, this is done from the total cell SS:

$$\text{Total cell } SS = [(T_0^2 + T_0^2 + T_0^2 + T_0^2 \ldots \ldots T_0^2)/n \text{ per cell}] - C,$$

where T_O is the total of each cell in the two-way matrix and df = number of cells − 1,

To compute the ABC × abc interaction, $SS_{A \times a}$ = total-cell $SS - SS_A - SS_a$.

The total-cell SS is calculated from the matrix and, SS_A and SS_A subtracted as shown below:

$$\text{Total-cell } SS = \frac{T_{aA}^2 + T_{aB}^2 + T_{aC}^2 + T_{bA}^2 + T_{bB}^2 + T_{bC}^2 + T_{cA}^2 + T_{cB}^2 + T_{cC}^2}{n \text{ per cell}} - C,$$

and cell df_a = the number of cells − 1.

The interaction sum of squares is then obtained by subtraction, where

$$SS_{A \times a} = \text{Cell total } SS - SS_A - SS_a$$

and the interaction df, $df_{A \times a}$, is likewise obtained by subtraction: $df_{A \times a}$ = Total-cell $df - df_A - df_a$, or by merely multiplying df for the factors involved in the interaction, or $df_{A \times a} = df_A \times df_a$.

To compute the interaction sum of squares, $SS_{A \times a}$ between ABC and $\alpha\beta$: This is done from the total cell SS:

$$\text{Total cell } SS = \frac{T_{\alpha A}^2 + T_{\alpha \beta}^2 + T_{\alpha C}^2 + T_{\beta A}^2 + T_{\beta B}^2 + T_{\beta C}^2}{n \text{ per cell}} - C,$$

and cell df_a = the number of cells − 1.

The $\alpha\beta$ × ABC interaction:

$SS_{\alpha \times A}$ = Cell total $SS - SS_\alpha - SS_A$ can be calculated from the matrix and, SS_α and SS_A subtracted and the interaction df, $df_{\alpha \times A}$, is likewise obtained by subtraction, $df_{\alpha \times A}$ = Total-cell $df - df_\alpha - df_A$, or by merely multiplying df for the factors involved in the interaction, or $df_{\alpha \times A} = df_\alpha \times df_A$.

Similarly, the interaction sum of squares, $SS_{a \times \alpha}$, between abc and αb; and the $df_{a \times \alpha}$ can be calculated as described above. At this point, we have the between-treatment effects, SS_A, SS_a, and SS_α, and the interaction effects, $SS_{A \times a}$, $SS_{A \times \alpha}$, and $SS_{a \times \alpha}$.

Three-Way Interactions.   The three-way interactions can be calculated as follows:

Total cell $SS = [(T_.^2 + T_.^2 + T_.^2 + \ldots T_.^2/n$ per cell)$] - C$, where $T.$ is the Total per cell in a three-way matrix and Total cell df = number of cells $- 1$. From the total-cell SS and df, the three-way interaction can be calculated: The (ABC) $\times$ (abc) $\times$ ($\alpha\beta$) interaction,

$$SS_{A \times a \times \alpha} = \text{Total-cell } SS - SS_A - SS_a - SS_\alpha - SS_{A \times a} - SS_{A \times \alpha} - SS_{a \times \alpha},$$

and $df_{A \times a \times \alpha} = \text{Total-cell } df - df_A - df_a - df_\alpha - df_{A \times a} - df_{A \times \alpha} - df_{a \times \alpha}$ or $= df_A \times df_a \times df_\alpha$.

Error.   The error is obtained by subtracting all terms from the total,

Error $SS_E = \text{Total-cell } SS - SS_A - SS_a - SS_\alpha - SS_{A \times a} - SS_{A \times \alpha} - SS_{a \times \alpha} - SS_{A \times a \times \alpha}$,

$df_E = \text{Total-cell } df - df_A - df_a - df_\alpha - df_{A \times a} - df_{A \times \alpha} - df_{a \times \alpha} - df_{A \times a \times \alpha}$.

$SS_E = SS_T - \text{Total-cell } SS$ (for the three-way matrix), $df_E = df_T - \text{Total-cell } df$ (for the three-way matrix).

Data Relationships, Correlation, and Regression

Correlation Analysis

Correlation analysis is one of the most commonly used statistical techniques to determine whether two variables are related. For example, correlation analysis can be useful in determining which descriptive attributes are related to overall liking. Correlation analysis is also useful in searching for corresponding consumer and descriptive analysis terms (Popper et al., 1997). The correlation coefficient, r, is a summary statistic, a single value that can be used to determine the degree and the significance of the relationship. Any time a correlation coefficient is calculated, it is a good idea to plot the data points in a scatter plot. A graph may show that the data changes in a curvilinear manner indicating a nonlinear relationship, even though the assumption made in calculating a correlation coefficient is a straight line or linear relationship (Jones, 1997). Often, a correlation analysis is erroneously interpreted as a cause-and-effect relationship. A high correlation coefficient and a low probability that the correlation occurred by chance does not establish a causal relationship (Jones, 1997). A correlation table is shown in Appendix E.6.

Pearson Product Moment Correlation

For most people this is known as the correlation coefficient and is usually represented by the symbol r. The basic hypothesis in calculating r is a linear relationship between the two variables: $Y = b_0 + b_1X + e$, where Y is the dependent variable, X is the independent variable, B_0 is the intercept, and b_1 is the slope (or coefficient) of X, and e is the difference between the observed Y and the true value of Y. Correlation coefficients (r) lie between the values of -1.00 and $+1.00$.

Regression Analysis

Regression analysis is usually used for predictive purposes to relate the independent variables to the dependent variables. When graphical representation of the data show a nonlinear response, regression using a polynomial or quadratic model may be appropriate. In regression analysis, the sums of the variables, their squares, and their cross products are obtained. The results of regression analysis are frequently reported in analysis of variance tables (Jones, 1997). F values are obtained, and from the value of F and appropriate statistical tables, one can determine the statistical significance of the regression and the significance of the coefficients (intercept, slopes, and interaction). A regression analysis also results in a coefficient of determination (R^2), or the variation in y explained by x. In univariate regression, R^2 is the square of the correlation coefficient (r).

Residuals

A plot of the residuals, the observed value of y minus the value of y predicted by the equation (y axis) versus the predicted value of y (x axis) should show a random distribution of points. A nonrandom distribution means that errors associated with poor fit of the regression may be due to a systematic effect, such as lack of a higher order term or a need to transform the data prior to generating an equation (Rothman, 1997).

Multiple Regression

In multiple regression, the equation $Y = b_0 + b_1X_1 + b_2X_2 + b_3X_3 + b_4X_4 \ldots b_nX_n + e$ is used where the subscripts refer to the individual independent variables X_i and their associated regression coefficients (b_i). Curvilinear relationships with one or more variables can be investigated and defined using equations such as $Y = b_0 + b_1X_1 + b_2X_2 + e$, $Y = b_0 + b_1X_1 + b_2X_2 + b_{12}X_1X_2 + e$. Note that the subscript 12 is used to denote the coefficient of the product of variable X_1 and X_2.

Validation

Validation is an important step in determining whether an equation is truly predictive. For large data sets, validation can be accomplished by using a subset of the data. For small data sets, validation can be carried out using a new data set (Rothman, 1997).

Multivariate Methods

Within the last two decades, there has been a remarkable expansion in the use of multivariate statistical methods to examine sensory data. Multivariate methods have played a major role where assessments of quality are made against the background of all the other qualities of the food (Powers, 1981). Multivariate analysis can help us understand the underlying properties that are measured in evaluating the quality of food and help establish which variables are determinants of food quality. Multivariate methods are better suited for studying the data relationships between consumer evaluations and descriptive analysis results or physicochemical measurements (Resurreccion, 1988).

Principal Components Analysis

Principal components analysis (PCA) is a method of extracting structure from the variance–covariance or correlation matrix (Federer, 1987). Its objective is the interpretation of data relationships (Popper et al., 1997). PCA constructs linear combinations of the original data with maximal variance. One uses PCA to reduce the number of variables represented to a smaller number of components, with little loss of information. A simple correlation table may show that several groups of variables are related to each other more closely than they are to any other variable or set of variables. To visualize this concept, it may be useful to think of the data as a cloud of points in space. The axis through which most of the points in the cloud lie is the first principal component. The equation of the axis will consist of those variables that do the most to reduce the variability of the cloud. The second principal component is the axis selected at right angles to the first axis that produces the maximum reduction in unexplained variation. The third axis is likewise at a right angle to the preceding axis and produced the maximum reduction in unexplained variation (Jones, 1997). In other words, the first component will account for the greatest portion of the variance, the second for the second largest portion, and so on, until all the variance has been accounted for. Each of the principal components is a linear combination of some of the original variables. As many components will be derived as necessary to account for the linear structure in the original variables (Resurreccion, 1988). In practice, the project leader determines the limit on the number of components formed. It is therefore possible to create two or three principal components that can represent in excess of 70% of the observed variation.

Examination of the original variables that are grouped in the principal components usually gives meaningful insight into the type of variation being explained by each of the principal components. In addition, these groupings of variables can be graphically presented to show product separation in two- or three-dimensional space that can be visualized. In many cases, the graphical representation of the results can be much more revealing than the numerical results alone (Jones, 1997).

Factor Analysis

Factor analysis (FA) is a technique that is most commonly used in food quality studies for data reduction and simplification. The method is used to reduce a large number of variables to a smaller set of new variables, called factors, that can be used to explain the variation in the data. The objective in FA is to find a smaller number of factors that together can replace the original variables measured in the study. Although the algorithms used in factor analysis differ from PCA, they are applied in similar ways to sensory data. It is not uncommon to start with a PCA to obtain some insights that can be used to initiate factor analysis (Jones, 1997), and in some instances, it may lead to similar results.

In factor analysis, the factors are obtained by algorithms that work with correlations of the variables as opposed to variances, which are commonly used in PCA. In many cases, the axes found by factor analyses are treated by a mathematical operation called "rotation." The rotated axis yield a better alignment with the original axes. It is therefore possible to make a clearer interpretation of the resulting pattern of data points.

Cluster Analysis

Cluster analysis (CA) is a general term for procedures that group variables or cases according to some measure of similarity (Ennis et al., 1982). CA uses a variety of mathematical and graphical tools to locate and define groupings of data. The variables within a cluster are highly associated with one another, while those in different clusters are relatively distinct from one another (Cardello and Maller, 1987). CA is primarily used for multivariate data and can be used to examine relationships either among variables or individuals (Jones, 1997). Although some clustering methods permit the use of nominal data, most methods require the data at least to be ordinal (Jones, 1997). Several clustering methods are used, and all of them operate by determining some measure of distance between observations or groups of observations, the most commonly used being Euclidean distance, which is analogous to simply measuring the distance with a ruler (Jones, 1997). Another common measure is one minus the correlation coefficient $(1 - r)$. Whatever measure is used, the computations assign individuals to a cluster so as to minimize the distances among the points in the cluster.

Discriminant Analysis

Discriminant analysis (DA) is a multivariate technique aimed at determining which set of variables best discriminates one group of objects from another (Resurreccion, 1988). This technique allows the researcher to classify a food product sample into one of several mutually exclusive groups on the basis of several variables simultaneously (Frank et al., 1990; Ward et al., 1995). The project leader is interested in understanding group differences or in predicting correct classification in a group based on the information on a set of variables or when probabilities of group membership must be determined. In food quality measurements, the predictor variables are either instrumental measurements or sensory attribute ratings from descriptive analysis tests, or both, and the food products are grouped into acceptable (hedonic score of six and above) or not acceptable (below six), as determined through a consumer acceptance test. In some ways, it is similar to regression analysis. A "training set" of data is fitted to a mathematical function that will give each observation the highest probability of being assigned to the known proper population while minimizing the probability that the same observation will be misclassified. The discriminant functions developed can then be used to predict the acceptability of a given sample through the calculation of posterior probabilities of membership to either the acceptable or unacceptable group (SAS Institute, 1985; Frank et al., 1990) through instrumental or sensory descriptive analysis intensity ratings.

It is possible that only a subset of the original set of variables may be needed to create a discriminant function. In most cases, it is thought that the classifications of discriminant analysis are useful only to determine the classification of a new data set. However, it is a means of learning how seemingly unrelated variables work together to describe and categorize products. It may be of interest to determine which of the sensory and instrumental variables, when used together, do the best job of distinguishing among several different products.

From such information, combinations of data can be obtained that will define the relationship of various products. This type of knowledge would allow tailoring a product to better compete in a specific market. Similarly, when a new product is developed, it could be determined whether there was a match with one or more of the products from which the discriminant function was generated. The sensory and instrumental data can be used to determine the closeness of the match by entering them into the function and finding the probabilities associated with the new product having come from each of the known populations (Jones, 1997).

Stepwise discriminant analysis (SDA) is sometimes used as a procedure to combine variables that have significant discriminating power for classification purposes, in other words, the best function with the fewest number of variables possible. One important problem for discriminant analysis must be pointed out. This is the fact that the mathematical techniques are impartial to the theoretical relevance of one variable over another (Cardello and Maller, 1987). The set of discriminating variables may not be the best (maximal combination). To secure a maximal solution, one would have to test all possible combinations (Klecka, 1980). If two variables are highly correlated and effect equal discrimination, only one will be selected. An important discriminator may not be selected at all (Resurreccion, 1988). Problems of colinearity among variables can be minimized by using common sense in interpreting the data from discriminant analysis and "forcing" in variables when necessary for theoretical or practical reasons (Cardello and Maller, 1987). When a variable is forced on the data by the project leader, it may result in an overlap between groups to an extent that the separation of groups is not that powerful (Sheth, 1970). A second problem concerns validation of the analysis. If the same sample is used, it overestimates the predictive power of the discriminant function. A suggestion would be to split the sample in half, using half for analysis and the other half for validation (Sheth, 1970). These problems have led researchers to believe that discriminant analysis should be used not so much for prediction as to determine the relative importance of predictor variables on the basis of the structural analysis of discriminate coefficients (Resurreccion, 1988).

Canonical Analysis

In this analysis, there are two sets of measurements that have to be correlated: a criterion set and a predictor set. The underlying principle is to create a linear combination of each set of variables for each sample by obtaining a set of weights that maximizes the correlation between the two sets. Canonical analysis is appropriate when the researcher is primarily interested in the overall relationship between the predictor and criterion sets. The researcher may then isolate those predictor variables that contribute most to this overall relation. In addition, he may note which of the criterion variables are more affected (Sheth, 1970).

Multiple Regression

This is a statistical technique that is essential for relating independent variables such as instrumental measurements or sensory descriptive analysis ratings to dependent variables such as consumer affective responses. As in discriminant analysis, a series of predictor variables is employed to predict some dependent variable. In food quality evaluation, the most common use of multiple regression is to predict the magnitude of consumer affective responses on the basis of a series of measurements (instrumental or sensory) of the food (Cardello and Maller, 1987).

Selecting the Best Method

There is no single method that is superior to others in establishing correlations between physicochemical measurements and sensory assessment of quality (Powers, 1981). The ultimate choice must be made by the individual investigator (Cardello and Maller, 1987). Powers (1981) has suggested the following protocol

1. Acquire basic information regarding significant relationships between physicochemical measurements and sensory responses. The purpose of this first stage is to get to a point where one has enough objective facts to proceed with further exploration. It is advisable to look at plots to better understand results.

2. Refine the data to permit fuller understanding of the instrumental, physicochemical, and sensory measurements. Researchers should watch out for outliers and variables or observations that do not belong.

3. In the final stage, validation trials should be carried out to determine if the answers have really been found. This can be performed by full prediction testing on new data sets or by a split-sample validation, repeating the same data analysis method, and each time keeping part of the data for prediction testing.

References

ASTM, Committee E-18. 1996. *Sensory Testing Methods.* ASTM Manual Series: MNL 26, 2nd ed. E. Chambers, IV and M. B. Wolf, eds. American Society for Testing and Materials, West Conshohocken, PA. pp. 93–98, 102–107.

Cardello, A. V., and Maller, O. 1987. Psychophysical bases for the assessment of food quality. In *Objective Methods in Food Quality Assessment,* J. G. Kapsalis, ed. CRC Press, Boca Raton, FL, p. 61.

Cochran, W. G., and Cox, G. M. 1957. *Experimental Design,* 2nd ed. Wiley, New York.

Ennis, D. M., Boelens, H., Haring, H., and Bowman, P. 1982. Multivariate analysis in sensory evaluation. *Food Technol.* 36(11):83.

Federer, W. T., McCulloch, C. E., and Miles-McDermott, N. J. 1987. Illustrative examples of principal component analysis. *J. Sensory Studies* 2(1):37.

Frank, J. S., Gillett, R. A. N., and Ware, G. O. 1990. Association of *Listeria* spp. Contamination in the dairy processing plant environment with the presence of staphylococci. *J. Food Protection* 53:928–932.

Gacula, M. C., Jr., and Singh, J. 1984. *Statistical Methods in Food and Consumer Research.* Academic Press, Orlando, FL.

Iversen, G. R., and Norpoth, H. 1987. *Analysis of Variance.* Sage Publications, Newbury Park, CA.

Jones, R. M. 1997. Statistical techniques for data relationships. In *Relating Consumer, Descriptive, and Laboratory Data,* A. Munoz, ed. ASTM, West Conshohocken, PA.

Klecka, W. R. 1980. Discriminant analysis. Sage University Paper Series on Quantitative Applications in the Social Sciences, 07-019. Sage Publications, Beverly Hills, CA and London.

Meilgaard, M., Civille, G. V., and Carr, B. T. 1988. *Sensory Evaluation Techniques.* CRC Press, Boca Raton, FL.

O'Mahony, M. 1982. Some assumptions and difficulties with common statistics for sensory analysis. *Food Technol.* 36(11):75–82.

O'Mahony, M. 1986. *Sensory Evaluation of Food.* Marcel Dekker, New York.

Petersen, R. G. 1986. *Design and Analysis of Experiments.* Marcel Dekker, New York.

Popper, R., Heymann, H., and Rossi, F. 1997. Three multivariate approaches to relating consumer to descriptive data. In *Relating Consumer, Descriptive, and Laboratory Data.* A. Munoz, ed. ASTM, West Conshohocken, PA. pp. 39–61.

Powers, J. J. 1981. Multivariate procedures in sensory research: Scope and limitations. *Tech. Quart. Mast. Brewers' Assn.* 18(1):11.

Resurreccion, A. V. A. 1988. Applications of multivariate methods in food quality evaluation. *Food Technol.* 42(11):128, 130, 132–134, 136.

Rothman, L. 1997. Relationships between consumer responses and analytical measurements. In *Relating Consumer, Descriptive, and Laboratory Data,* A. Munoz, ed. ASTM, West Conshohocken, PA. pp. 62–77.

SAS Institute, 1985. SAS User's Guide: Statistics, SAS Institute, Cary, NC.

Sheth, J. N. 1970. Multivariate analysis in Marketing. *J. Advertising Res.* 10(1):29.

Simon, P. W., Peterson, C. E., and Lindsay, R. C. 1980. Correlations between sensory and objective parameters of carrot flavor. *J. Agric. Food Chem.* 28:559.

Snedecor, G. W., and Cochran, W. G. 1980. *Statistical Methods,* 7th ed. Iowa State University Press, Ames, IA.

Stone, H., and Sidel, J. L. 1985. *Sensory Evaluation Practices.* Academic Press, San Diego, CA.

Stone, H., and Sidel, J. L. 1993. *Sensory Evaluation Practices,* 2nd ed. Academic Press, San Diego, CA.

Ward, C. D. W., Resurreccion, A. V. A., and Mcwatters, K. H. 1995. A systematic approach to prediction of snack chip acceptability utilizing discriminant functions based on instrumental measurements. *J. Sens. Stud.* 10:181–201.

13

Quantification of Quality Attributes as Perceived by the Consumer

Introduction

The perception of quality characteristics (both external and internal) determines a consumer's decision to purchase and repurchase a product (IFT, 1990). Before quality can be evaluated, it should be defined. The definition of the sensory quality of a food or food product is "the acceptance of the sensory characteristics of a product by consumers who are regular users of the product category, or who comprise the target market for the product" (Galvez and Resurreccion, 1992).

Dr. Roland Harper (1981) developed a scheme that specifies the interrelationships between the affective value or hedonic value of a food and its sensory and physicochemical attributes. In this scheme, physicochemical variables are related to sensory variables in a psychophysical relationship between stimulus and sensations, and there are relationships between these sensations and hedonic variables. This relationship is called *psychohedonic* because there are two types of psychological variables. It was noted that efforts to explain sensory acceptance on the basis of physicochemical variables directly have not been very successful so far because the sensory variables are often ignored (Frijters, 1988).

Strategic Approach

A strategic approach to the development of a food product is shown in Figure 13.1. Much of the success or failure of a food product in the marketplace is a result of consumers' perception of sensory quality. The strategic approach to product development involves, as the very first step, the identification of a product, then the identification of critical quality attributes of the product by regular consumers of a product.

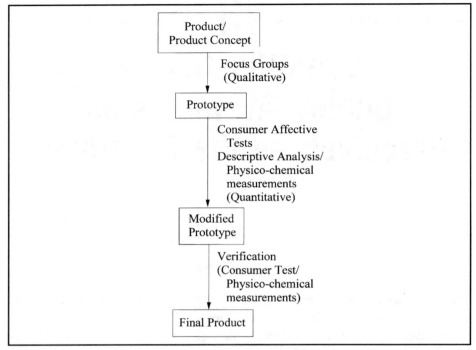

Figure 13.1   A strategic approach to the development of a food product (Resurreccion, 1997; reprinted with permission).

Concept Stage

Product concepts may arise from a variety of sources, including the marketing and research departments, management input, consumer research, and idea generation. Only after the quality attributes that are important to consumers have been determined can systematic product optimization take place. Qualitative research in the form of focus groups is used to identify the critical quality attributes.

Qualitative Research

Focus groups consist of 8–10 consumers recruited to fit specific demographic, attitudinal, and usage characteristics; they are conducted by a trained moderator, who helps to stimulate and direct the discussion (ASTM, 1979; Sokolow, 1988). The focus group is particularly valuable in language generation and product development research (Sokolow, 1988). With this information, formulation or process variables can be systematically modified to increase the likelihood of acceptance of the final product by consumers. Focus groups are an effective means to identify those attributes that are important in the product and should be included and optimized, and those characteristics that are undesirable to consumers and should be minimized or eliminated from the product. Focus group results can be used to assist in questionnaire design of the succeeding consumer test. A detailed discussion of focus groups is in Chapter 5.

Quantification of Product Quality

Quantification occurs after the critical quality attributes have been identified. Beyond concept development, initial prototype products are developed that possess the critical attributes valued by the consumers and that minimize or are devoid of the attributes consumers consider unacceptable. At this stage, consumer sensory tests are a useful tool for product guidance. The consumer tests help to assess performance potential and provide guidance for further development (Resurreccion, 1997). Results of the affective tests may result in modification of the prototypes.

Verification of Consumer Acceptance

Additional consumer affective tests are conducted to quantify overall acceptance and consumer responses to the critical attributes that determine the product's acceptance. A significant positive response may lead to the development of a final product. Descriptive analysis and physicochemical measurements are used to characterize and establish quantitative limits for these critical attributes. A successful acceptance test will not guarantee success of the product in the marketplace and is not a substitute for large-scale market tests.

Final Product

At this point, execution of the test market research is the function of the market research department. Consumer tests will be necessary as production capabilities are increased. The effect of formulation or process changes associated with scale-up needs to be determined at this stage to ensure that these changes do not compromise consumer acceptance (Resurreccion, 1997).

Acceptance and Preference Tests

Acceptance tests measure acceptability or liking for a food by consumers. Preference tests measure the appeal of one food or food product over another (Stone and Sidel, 1993). The methods most frequently used to determine preference and quantify acceptance are the paired-preference tests and tests employing the 9-point hedonic scale, respectively. Ratio scaling was proposed by Moskowitz (1974) as a method to quantify acceptance and preference but is not used as frequently as the paired-preference tests and tests employing the 9-point hedonic scale.

The 9-point hedonic scale has been used for a number of years and is validated in the scientific literature (Sidel and Stone, 1993). However, in some instances such as in testing children's responses, adaptations of the 9-point hedonic scale in the form of a 9-point facial scale were useful for children eight years old or older (Kroll, 1990; Kimmel et al., 1994). The 3-, 5-, and 7-point facial hedonic scales were found to be more appropriate for three, four, and five-year-olds, respectively (Chen et al., 1996).

Consumer Affective Test Methods

Consumer responses needed for the quantification of acceptance or preference can be conducted in a sensory laboratory setting, in central location tests (CLTs), or in home-use tests (HUTs), which are also known as home placement tests. A specialized form of CLT that has been used is one that is conducted in a mobile laboratory.

Sensory Laboratory Tests

Sensory laboratory tests, as the name implies, are conducted in the sensory laboratory of a food company, consulting firm, or research organization. This type of test is the most frequently used method for obtaining consumer responses. A more complete description of the sensory laboratory test is found in Chapter 6.

The panel consists of consumers who are recruited from a consumer database consisting of prerecruited consumers screened for eligibility to participate in the tests. Usually, 25 to 50 responses are obtained; however, 50 to 100 responses are considered adequate (IFT/SED, 1981). The recommended number of products per sitting is two to five.

The major advantage of using the sensory laboratory location to conduct a consumer affective test is the convenience of the location. The sensory laboratory is usually located within the building of the food company, consulting firm, or research organization. The sensory laboratory provides a convenient location for the research team but is less convenient for consumer panelists. If the participants in the affective test have previously participated in consumer tests, they are familiar with the testing procedures, thus saving the researcher time. In the sensory laboratory, the researchers are able to control the conditions of the testing environment such as lighting, noise, and other distractions, and to conduct the test in individual, partitioned booths where panelists are isolated from each other and therefore do not influence each other. Furthermore, because the sensory laboratory is usually constructed adjacent to a fully equipped kitchen, sample-preparation steps including cooking, reheating, slicing, and serving can be conducted under highly controlled conditions. Another advantage of conducting a consumer test in the sensory laboratory is the rapid turnaround time for results to be obtained because of the proximity to the data processing facilities.

The disadvantages of conducting consumer tests in the sensory laboratory are the limitations on the consumer panel being used. If consumers screened from a prerecruited consumer database are used, those consumers who would be most willing to participate are those whose homes or place of employment are close enough to the sensory laboratory. If care is not exercised to recruit an appropriate panel, the demographic characteristics of the panel may be skewed on characteristics such as income, education, race, and any other factors that may have an influence on usage patterns for the food. Another disadvantage is that the temptation to use company employees is great. The project leader needs to remember that there is some risk associated with using employees in product maintenance tests; employees should not be used in product development, product improvement, or product optimization tests. If a decision to use employees in product maintenance is made, care must be taken to recruit only those employees who are not familiar with the production, testing, or marketing of the product; in addition, their rating patterns must be compared to those of a representative panel. A method to determine whether an employee panel would present great risk and should not be used is discussed in Chapter 4.

Central Location Tests

CLTs are conducted in a location away from the sensory laboratory that is accessible to the public, such as a shopping mall, hotel, or food-service establishment. A discussion on the central location test is in Chapter 7.

The panel may consist of consumers who are recruited from a consumer database consisting of prerecruited consumers, but more frequently, consumers are intercepted to participate in the tests, screened for eligibility, and, if qualified, are immediately recruited to participate in the tests. Usually 100 or more responses are obtained. Several central locations may be used in the evaluation of a product. The recommended number of products to be evaluated per sitting is one to four.

The major advantage of the CLT is that the location offers the ability for sensory research personnel to recruit and test a large number of participants.

In addition, actual consumers participate in the test; no company employees are used. The disadvantage of the CLT is the distance from the food company, consulting firm, or research organization. As a result, the central location often has limited facilities, equipment, and resources for food preparation and conduct of the test. In addition, facilities that may be lacking in a central location are space for a registration and orientation area, kitchen and food-preparation equipment, and individual partitioned booths for the sensory test. The inadequacy of testing facilities would limit the type of tasks that can be performed by panelists and may pose a severe limitation on the ability to conduct the test under the experimentally controlled conditions required. The large number of panelists that can be recruited for this type of test is a disadvantage as much as it is an advantage; the location is ideal for recruitment of a large panel, but accompanying the large panel size are the time and manpower requirements needed to conduct the test. Data are collected by trained interviewers rather than by self-administered questionnaire, adding to the number of personnel needed for the test and the time required to collect data from a given number of respondents.

Mobile Laboratory Tests

A special form of central location test is a hybrid between the sensory laboratory test and the central location test. A fully equipped mobile laboratory with complete facilities for food preparation and sensory testing is driven to and parked at a central location, and the tests are conducted within the mobile laboratory. The mobile laboratory test is described in Chapter 8.

The panel consists of consumers who are intercepted to participate in the tests. There is usually no prerecruitment or screening except for age. Usually, 75 or more responses are obtained in one location, and the mobile laboratory is usually driven to other locations, where additional tests are conducted. Several central locations may be used in the evaluation of a product. Data are collected by trained interviewers rather than by self-administered questionnaires, adding to the number of personnel needed for the test and the time required to collect data from a given number of respondents. The recommended number of products to be evaluated is two to four.

The major advantage of using the mobile laboratory for a consumer affective test is the ability to recruit and use a large number of "actual" consumers for the test and have the facilities to maintain experimentally controlled facilities and environmental conditions for food preparation and testing. The disadvantages of the mobile laboratory test are the expense of maintaining the laboratory and logistical arrangements that have to be made prior to the test for parking and power supply.

Home-Use Tests

HUTs are also referred to as home placement tests. These tests, as the name implies, require that the test be conducted in the participants' own homes. Samples are either mailed to the consumers or consumers are asked to pick these up from the sensory laboratory or at a central location.

The panel consists of consumers who are usually prerecruited and are part of a consumer database, and are screened for their eligibility to participate in the tests. However, in practice, employees of a company who have little or no responsibility for production, testing, or marketing of a product are often used. The project leader needs to exercise caution when using employee participants in HUTs. There is some risk associated with these panels in product maintenance tests, and they should not be used in product development, improvement, and optimization experiments. Usually, over 100 responses are obtained per product. The number of products to be tested should be limited to one or two. Chapter 9 contains a more comprehensive discussion on the HUT.

The major advantage of the HUT is that the products are tested under actual home-use conditions. In addition, one is not only able to test a product for acceptance or preference but also for product performance. Furthermore, because the products are tested in the actual home environment, responses may be obtained not only from the respondent, who is usually the major shopper and purchaser of food in the household, but also from the other members of the family, or from the entire household. In many instances, other pieces of information may be obtained, such as the types of competitive products found in the home during the test, usage patterns of all household members, and other information that would be useful in marketing the product. If the participants of the tests are prerecruited and screened from an existing consumer database, the participants awareness of their role and the importance of the data collected will likely result in a high response rate.

The main disadvantages of the HUT are that it requires a considerable amount of time to implement, distribute samples to participants, and collect participant's responses. Little can be done to exert any control over the testing conditions once the product is placed in the home of the respondent. The HUT can be expensive in terms of product cost because sample sizes are much larger than one would serve in a sensory laboratory or CLT, although others would argue that the larger amounts served allow the food to be tested under actual home conditions, such as drinking rather than sipping a beverage, which would be the case in a laboratory, central location, or mobile laboratory test. The mailing or distribution of the products to participants may be a major expense associated with this test and will add to the cost of the test. If participants have not been prerecruited and screened from an existing consumer database, the participants will likely be less aware of the importance of their role and the importance of the data being collected. In such cases, response rates will be lower than expected.

Product Characterization

Once consumer acceptance for the product has been quantified, the sensory practitioner defines the lower boundary for consumer acceptance, with input from company management. Some companies that pride themselves on the quality of their products may be unwilling to produce a product that is only "liked slightly" (= 6) on a 9-point hedonic scale and may opt to accept a higher limit set at "like moderately or higher" (= 7) for the product. Simultaneously, characterization of the product tested using consumer affective tests may take place using various established instrumental or physicochemical methods, or sensory descriptive analysis tests.

Physicochemical Measurements of Product Quality

Product characterization may be conducted using sensory descriptive analysis tests when no instrumental or physicochemical test can be used to validly and reliably characterize an attribute. If an instrumental test is available that accurately characterizes the sensory attributes of a product, it makes little sense to use a descriptive sensory panel to quantify the attribute. Physicochemical measurements have been used to characterize attributes such as color, flavor, tastes, texture, and viscosity.

Instrumental color measurements of lightness, chroma, and hue angle can be obtained using a colorimeter. Texture can be quantified using measurements of shear, cutting force, work to cut, or parameters calculated from instrumental texture profile analysis using an Instron universal testing machine or texture analyzer. Examples of texture measurements are discussed later. A number of flavor attributes can be calculated from prediction equations using peak areas from gas chromatograms of headspace of known volatile compounds. Tastes can be quantified by analysis of chloride, sucrose, or acid. Examples of some of these physicochemical measurements are described in the following sections.

Color

Appearance may be the single most important sensory attribute of food. Decisions of whether to purchase and/or ingest a food are determined in great part by appearance. Color has also been shown to influence perception of sweetness (Clydesdale, 1991; 1993) and flavor intensity (Clydesdale, 1993; Christensen,

1983). Tristimulus colorimetric methods that are highly correlated to human perception of color have been developed (Clydesdale, 1978).

Texture Profile Analysis

Success of a product is dependent on taste, followed by texture and mouthfeel (Duxbury, 1988). The Instron universal testing machine can be used to characterize textural properties of foods. Texture profile analysis (TPA) can be made using the Instron fitted with a compression cell with two cycles of deformation. The two cycles simulate the chewing action of teeth (Bourne, 1982). The properties of fracture, hardness, springiness, and cohesiveness can be calculated from the force-deformation curves resulting from TPA. TPA curves have been generated on several products including pretzels and bread sticks (Bourne, 1982).

Kramer Shear-Compression Test

The Kramer shear-compression test has been used for texture investigation of crisp foods such as potato chips, crunch twists, and saltine crackers (Seymour and Hamann, 1988). One cycle of deformation is used and the parameters of maximum force at failure, work done at failure, and force to shear are used to describe the food's texture attributes (Seymour and Hamann, 1988; Bhattacharya et al., 1986).

Snapping Test

Sensory quality attributes such as crispness have been related to instrumental texture measurements. Kramer shear-compression test information is difficult to interpret and does not always mimic the action of biting into a sample for crispy foods. A 1mm blunt blade attached to the Instron with the sample placed on two bars sufficiently apart to prohibit friction during the deformation cycle can be used to simulate the process. This test is called the snapping test. It has been used on crisp foods such as crackers (Katz and Labuza, 1981); crisp bread, potato crisps, ginger biscuits (Sherman and Deghaidy, 1978); snap cookies (Bruns and Bourne, 1975; Curley and Hoseney, 1984); and cream cracker biscuits (Gormley, 1987). The slope of the deformation curve is related to crispness and fracturability.

Descriptive Analysis Measurements

There are few sensory attributes that can be accurately predicted by instrumental and physicochemical measurements. In such cases, sensory descriptive analysis techniques are used to characterize critical attributes. Sensory descriptive analysis testing is a widely used technique in the sensory analysis of food materials to characterize products when there are no instrumental or physicochemical tests that can be used to validly and reliably characterize product attributes. Descriptive analysis is a sensory method by which attributes of a food or food product are identified and quantified using human assessors specifically trained for this sensory method. Panelists are selected on their ability to perceive differences between test products and verbalize perceptions (IFT/SED, 1981). The panel of assessors is used as an analytical tool in the laboratory and is expected to perform like one (O'Mahony, 1991; Munoz et al., 1992). There are several standard techniques available, such as the flavor profile analysis (Cairncross and Sjostrom, 1950; Caul, 1957), quantitative descriptive analysis (Stone et al., 1974), spectrum descriptive analysis (Meilgaard et al., 1987), and the texture profile analysis (Brandt et al., 1963; Szczesniak et al., 1963).

Flavor Profile

The flavor profile is a method that describes the aroma and flavor of products or ingredients. The method consists of formal procedures for describing and assessing the flavor of a product in a reproducible manner. It involves a minimum of four assessors selected on the basis of screening tests, a personal interview to determine availability and personality traits, and training over a period of six months. The assessors evaluate the food or food product in a quiet, well-lit, odor-free room. It is suggested that a round table be used to facilitate discussion. Assessments take approximately fifteen minutes per sample. One to three sessions are held. Assessors make independent evaluations, rate character, and note intensities using a 7-point scale from threshold to strong (Resurreccion, 1997).

Quantitative Descriptive Analysis

Quantitative descriptive analysis (QDA) is a descriptive analysis method that describes all sensory properties of a product and quantifies their intensities. It is led by a descriptive analysis moderator and requires 10–12 panelists, although in some tests, 8–15 panelists may be involved. Training is best conducted in a conference-style room. During training, panelists develop terminology, definitions, and a standardized evaluation procedure. The training requires two weeks of training, or approximately eight to ten hours organized into ninety-minute sessions. The panel leader organizes the training, introduces the products, introduces references when needed, and facilitates panel activities, but does not participate in the evaluation of samples as an assessor. Products are evaluated in partitioned booths. Assessors take three to twenty minutes per product using a repeated trials design. A minimum of three replications is recommended.

References are provided as needed, and a minimum of three replications are recommended. Fifteen-centimeter line scales are used in rating samples (Resurreccion, 1997).

Spectrum Descriptive Analysis Method

The spectrum method utilizes a universal scale that is consistent for all attributes and products. Perceived intensities are rated according to these universal scales. The training is led by a moderator trained in the spectrum descriptive analysis method. Twelve to 15 trained assessors are required. Assessors are trained in a room around a table. Assessors develop the terminology, definitions, and evaluation techniques, and agree on references to be used during the evaluations. Training takes place over a period of three to four months. The trained and calibrated panelists evaluate food in partitioned booths. Evaluations are conducted in duplicate or triplicate in separate evaluation sessions. Evaluation is approximately fifteen min per product. Fifteen-point or 150 mm line scales are used (Resurreccion, 1997).

Texture Profile Analysis

A moderator trained in texture profile analysis leads texture profile analysis. Assessors gather around a table to facilitate discussion and evaluation. Assessors are trained on specific references for each specific attribute scale, texture definitions, and evaluation procedures. The way the product is manipulated in the mouth during the evaluation is standardized. Evaluation takes five to fifteen minutes per product; a minimum of three replications is recommended. References are provided as needed (Resurreccion, 1997).

Hybrid Descriptive Analysis Methods

Descriptive analysis methods that employ a combination of some aspects of both QDA and the spectrum analysis methods are used by a large proportion of sensory practitioners. These are called hybrid methods and are widely used in sensory descriptive analysis.

Modeling of Quality or Consumer Acceptance

Graphical Methods

In setting quality specifications, Munoz et al. (1992) describe a method to determine the nature of relationships between consumer acceptance and descriptive panel responses. Graphical methods are utilized, where overall acceptance

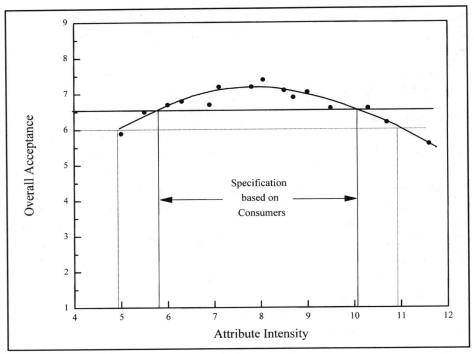

Figure 13.2   Establishment of specifications based on consumer response. Over-all acceptance was measured using a consumer panel that rated product on a 9-point hedonic scale (1 = "dislike extremely"; 5 = "neither like nor dislike"; 9 = "like extremely"). Attribute intensity was determined from mean ratings by a descriptive panel using a 15-point (150-mm) line scale (Munoz et al., 1992; reprinted with permission).

ratings or attribute acceptance ratings are plotted against attribute intensities for one variable in individual graphs. Overall consumer acceptance of sensory attributes of color, flavor, and texture can be plotted against color, surface appearance, flavor, or texture attribute intensities obtained from descriptive analysis measurements or instrumental and physicochemical measurements. The relationships found, if any, may be linear or curvilinear. Cardello et al. (1982), in studying relationships between judgments of perceived texture of frozen fish by trained and consumer panelists, observed good linear correlations between scalar judgments of texture, although a broader perceptual range was seen in trained panelists' responses. In another example (see Figure 13.2), a curvilinear relationship was found between overall acceptance and attribute intensity. When a decision on the lower limit of product acceptance is made by management from consumer acceptance test results, the product attribute

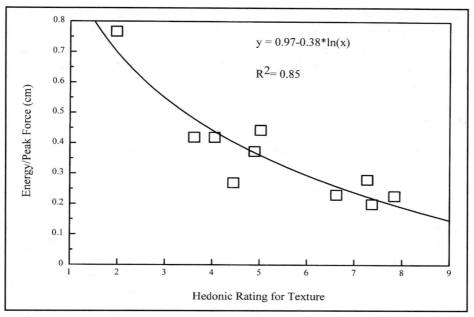

Figure 13.3 The relationship of energy to peak force (E/PkF) as measured by Kramer shear versus the overall texture acceptability of commercial and experimental snack chips measured on a hedonic scale (1 = "dislike extremely"; 5 = "neither like nor dislike"; 9 = "like extremely") (Ward, 1995).

specifications can be determined from plots similar to Figure 13.3. Although this type of plot may be applicable to a wide variety of products and product attributes, let us select the attribute hardness. In this example, the product was liked best and rated a 7, or "like moderately," on a 9-point scale, and the corresponding product was rated during texture profile analysis approximately 7.8 on a 15-point scale for hardness. The graph indicates that the product will be liked by consumers (minimum overall acceptance rating of 6) as long as the hardness intensity rating is between 4.95 and 10.9. Overall acceptance decreases with an increase or decrease in hardness. The company specification for product quality will be based on management's decision of the level of overall acceptance of the product that they would like to have. A number of companies may decide that they will accept an overall acceptance of 6.5. In this case, product hardness should be between 5.7 and 10.1 on a hardness scale. Should management decide to relax their standards and go with a product with an overall acceptance of 6, hardness should range from 4.95 to 10.9. On the other hand, some companies may pride themselves on a high-quality product that has a mean rating of 7. In this case, the product's hardness intensity specification is narrowed to 6.6 to 9 on the 15-point scale. This procedure, used for quantifying overall acceptance one variable at a time, requires the graphing of overall acceptance against a number of descriptive attribute intensities or instrumental and physicochemical measurements, and can be tedious and time consuming.

Predictive Equations

Mathematical models may be developed using regression model such as

$$\text{Acceptance rating} = B_0 + B_1(X_1)$$

where the acceptance score is the dependent variable. B_0 is the intercept, B_1 is the regression coefficient, and X_1 is the independent variable. The independent variable may be the descriptive analysis ratings or instrumental and physicochemical measurements. A predictive equation for consumer ratings for juiciness of a meat product rated on a 9-point hedonic scale (Munoz and Chambers, 1993) was based on a descriptive analysis ratings from 0 to 150 for moisture release where

$$\text{Juiciness rating} = -0.658 + 0.098 \text{ (moisture release)}$$

A coefficient of determination (R^2) of .90 was obtained. In the next example, illustrated in Figure 13.3, a negative curvilinear relationship was found between hedonic ratings for texture of snack chips and the energy/peak force needed to break a chip as determined by a snapping test using the Instron universal testing machine (Ward, 1995). The higher the energy/peak force to shear the snack chips, the lower is the consumer texture acceptance rating. Conversely, the lower the energy/peak force, the higher is the overall texture rating. The acceptance of the texture of snack chips can be related to energy/peak force to shear the snack chips using the equation:

$$y = -0.38* \ ln(x) + 0.97$$

where y = acceptance of texture, and x = energy/peak force (Ward, 1995). This equation can be used to calculate the effect of energy/peak force on texture ratings; a high energy/peak force results in unacceptable texture in snack chip products and a low energy to peak force results in acceptable texture in snack chips. From this equation, an acceptance score of 7 ("like moderately") or higher may be obtained in samples with an energy/peak force of 0.25 cm to 0.15 cm. If a product texture acceptance score of 6 ("like slightly") is decided by management for the chip, the energy/peak force specification can be set accordingly at 0.29 cm to 0.15.

Response Surface Methodology

When the effect of two or more variables on product acceptance and its interrelationship with sensory attribute intensities is of interest, response surface methodology (RSM) simplifies the process. RSM is a designed regression analysis meant to predict the value of a response variable, or dependent variable, based on the controlled values of the experimental factors, or independent variables (Meilgaard et al., 1991). An experimental design is used that allows for the study of two or more variables simultaneously. The samples are evaluated by a consumer and descriptive panel, and regression analysis results in predictive equations. From the parameter estimates, it can be determined which variable contributes the most to the prediction model, thereby allowing the product researcher to focus on the variables that are most important to the product acceptance (Schutz, 1983).

The dependent variable is the acceptance rating and is the only rating that is absolutely necessary to obtain optimal formulations. From attributes intensities determined during descriptive testing, formulations that will result in acceptable products can be determined. Contour plots of the prediction models allow the researcher to determine the predicted value of the response at any point inside the experimental region without requiring that a sample be prepared at the point (Meilgaard et al., 1991).

The applications of RSM in a wide variety of areas, including food research, were reported by Hill and Hunter (1966). It has constantly been successfully demonstrated that it can be used in optimizing ingredients (Henselman et al., 1974; Johnson and Zabik, 1981; Vaisey-Genser et al., 1987; Chow et al., 1988; Shelke et al., 1990) and process variables (Oh et al., 1985; Floros and Chinnan, 1988; Mudahar et al., 1990; Galvez et al., 1990; Vainionpaa, 1991) or both (Bastos et al., 1991).

The following example demonstrates the use of RSM to define product quality. The objective of this study was to investigate the use of RSM to develop specifications for an acceptable roasted, salted peanut with decreased oil content.

Experimental Design

Three factors: percent salt, oil content, and degree of roast (color lightness, L, value) at three levels each were studied as follows: average degrees of roast were represented by Hunter color lightness (L) values of 45, 50, and 55. Salt levels were 1.5, 2.0, and 2.5 percent by weight, and average oil levels were

24%, 34%, and 47% by weight. The oil levels were not equidistant; the lowest level was 10% rather than 13% below the midpoint level due to the difficulty in further removal of oil in the peanuts by mechanical pressing. The full-factorial design resulted in 27 treatments (3 oil levels × 3 degrees or roast × 3 salt levels).

Descriptive Tests

Panelists evaluated 27 roasted, salted, defatted peanut samples prepared in three processing replications. One processing replication was evaluated each day over three days in a complete, balanced block design. Attribute intensities were rated using continuous line scales (Stone, 1992) in a computer scoresheet. Cluster analysis was used to evaluate performance of trained panelists and determine outliers (Malundo and Resurreccion, 1992), whose results were deleted from the data prior to analysis.

Consumer Test

A central location test was conducted using 27 treatments and a replication with a block size of 54 (SAS Institute, 1980). Design parameters incorporated 13 consumer responses per treatment per block, for a total of 26 responses per treatment between two processing replications. In a balanced, incomplete block design, each consumer evaluated 5 of 27 samples from one replication to determine a hedonic score for overall acceptance, and acceptability of color, appearance, flavor, and texture. A total of 142 consumers were required for the study.

The majority of participating consumers were females, as women were the primary household shoppers and there was no quota set for gender. The median age range was thirty-five to forty-four, and median household income was in the $40,000–49,000 range. The race distribution was 73.2% white and 26.8% black and other races. Results also indicated that 50% of panelists consumed snacks one to three times per day.

Statistical Techniques

After responses of one panelist found to be an outlier were deleted, scores of the remaining ten panelists were analyzed as follows:

Consumer and Descriptive Tests

A quadratic response surface model was fitted to the data using response surface regression analysis to determine the behavior of the response variables in relation to the independent variables fat, degree of roast, and salt studied (Freund and Littell, 1986). The model included all linear and quadratic terms in the individual independent variables and all pairwise cross products of linear terms. Not all of the variables in the full models from the previously mentioned model were significant in contributing to the predictive power of the models. A subset of variables determined to contribute significantly to the models was selected. Response variables that resulted in significant regression models at $p < .05$ and a coefficient of determination (R^2) greater than .20 were selected as contributing to, and included, in the predictive model. When a significant interaction term involved an insignificant linear term, the linear term was retained in the predictive model. Multiple regression analysis was used to finalize the models after significant variables were selected according to the procedure described above.

Contour Plots for Consumer Acceptance

Using the Surfer (version 4.15, Golden Software, Inc., Golden, CO) access system program, contour maps were plotted for each model where the independent variables were found to have significant effects on overall acceptance, and acceptance of appearance, color, flavor, and texture. In this study, the limit for product acceptance was set at a rating of at least 6 (= "like slightly") or higher on a 9-point hedonic scale for all attributes.

Prediction Equations for Attribute Intensities

Minimum levels for oil for light (L = 56) to dark (L = 44) degrees of roast, and percent salt (held constant at 2.0) for a product with an acceptable overall quality of 6 or greater, as determined from the contour plots, were substituted into prediction equations to determine the provided comparisons for a reduced-fat product with one-third less oil content; prediction equations were used to determine the attribute intensity range of a reduced fat product with acceptable quality.

Results

Response surface regression analysis of the consumer test data resulted in significant regression models at $p < .05$ for overall liking, color, appearance, flavor, and texture. Parameter estimates and coefficients of determination from acceptance scores used in prediction models are presented in Table 13.1. Salt level was not significant in any of the consumer acceptance models, and was not used in the predictive models. All coefficients of determination (R^2) were greater than .20; therefore, all acceptance attributes were considered in determining the quality to roasted, salted, defatted peanuts. Parameter estimates indicated the degree of roast and oil level as important factors affecting acceptance scores;

Table 13.1 Parameter Estimates and Coefficients of Determination (R^2) for Factors Used in Prediction Models for Consumer Acceptance of Roasted, Salted, Defatted Peanuts ($n = 142$)[a]

FACTORS	PARAMETER ESTIMATES				
	Overall liking	Color	Appearance	Flavor	Texture
Intercept	−21.23	−17.72	−15.49	−60.32	−17.14
Oil	0.63	0.54	0.49	0.709	0.52
Roast	0.45	0.40	0.33	1.934	0.35
Oil × Roast	−0.01	−0.0089	−0.0074	−0.0014	−0.0077
Roast × Roast	—	—	—	−0.014	—
R^2	0.2837	0.2297	0.2988	0.3249	0.3067

[a] Factors are oil = oil content of full fat or defatted peanut, roast = degree of roast (color lightness, L). Percent salt was not a significant factor in predicting acceptance. Panelists evaluated samples using a 9-point hedonic scale where 1 = "dislike extremely" and 9 = "like extremely."

however, the low regression coefficients for the interaction term between oil level and degree of roast were important in the prediction of acceptance scores.

The contour plots presented in Figure 13.4 were generated using the prediction models for overall acceptance, and acceptance of color, appearance, flavor, and texture. All combinations of oil and degree of roast on the plots that would result in products with acceptance ratings equal to or greater than 6, or "like slightly," are shaded. If scores corresponding to 6 or higher for overall liking (Figure 13.4E) alone are used to define a new product, all products with combinations of oil level from 48% to 39.1% oil and a degree of roast (color lightness, L) from 44 to 56, are shaded and will be acceptable. In addition, all combinations in the shaded area between 29.2 and 39.1% oil and an L value of 44 to 56 will yield acceptable products.

Parameters estimates and coefficients of determination for factors and variables used in attribute intensity prediction models for the roasted, salted, defatted peanuts are presented in Table 13.2. Variables from the full regression models contributed significantly at $p < .05$ to eight predictive models, for brown color, roast peanutty, raw/beany, burnt, astringent, oxidized, saltiness, and bitterness attributes.

The values of the parameter estimates indicated that the degree of roast and oil level were important factors affecting most of the descriptive scores. The level of oil was considered the most important factor (largest regression coefficients) affecting descriptive scores in brown color and raw/beany attributes, and degree of roast (color lightness, L) was considered the most important factor in roasted peanutty and burnt attributes. The parameter estimate for burnt

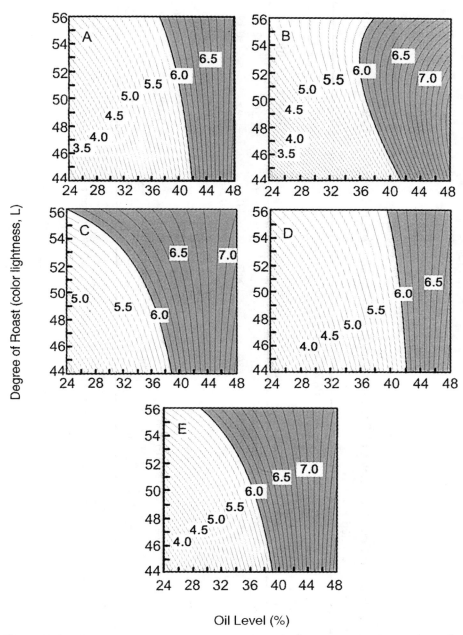

Figure 13.4 Contour plots for prediction models for consumer acceptance for treatments with varying oil levels and level of roast values for A. texture, B. flavor, C. color, D. appearance, and E. overall liking. Scores are based on a 9-point hedonic scale with 1 = "dislike extremely", 5 = "neither like nor dislike", and 9 = "like extremely." Shaded regions represent a hedonic acceptance score of 6 (= "like slightly") or greater for each attribute (Plemmons, 1997).

Table 13.2 Parameter Estimates and Coefficients of Determination (R^2) for Factors Used in Prediction Models for Descriptive Analysis of Roasted, Salted, Defatted Peanuts[a]

FACTORS	PARAMETER ESTIMATES							
	Brown Color	Roasted Peanutty	Raw/ Beany	Burnt	Astringent	Oxidized	Saltiness	Bitterness
Intercept	10.16	30.18	43.96	330.15	57.98	48.16	36.20	42.02
Oil	6.32	0.111	-1.78	-0.56	-0.17	-0.25	-0.65	-0.19
Roast	0.33	0.64	-0.22	-9.80	-0.44	-0.20	—	-0.33
Salt	—	—	—	-4.086	—	—	—	-0.72
Oil × Roast	-0.074	-0.037	0.019	—	—	—	—	—
Roast × Salt	—	—	—	0.08	—	—	—	0.026
Oil × Oil	-0.042	0.031	0.015	—	—	—	0.019	—
Roast × Roast	—	—	—	0.083	—	—	—	—
R^2	0.5463	0.2412	0.2325	0.3247	0.0993	0.0976	0.1874	0.1920

[a]Regression models for each attribute were significant at $p < .05$.

Factors are oil = oil content of full fat or defatted peanut, roast = degree of roast (color lightness, L), salt = percent salt.

Table 13.3 Predicted Attribute Scores from Descriptive Analysis Repression Equations for Characterization of Roasted, Salted, Defatted Peanuts of Acceptable Quality[1]

Attributes	Factor[2]	Sample[3]		
		A	B	C
	X_1	39.1	29.2	32.0
	X_2	44	56	54
	X_3	2.0	2.0	2.0
Brown color		81.0	56.8	59.9
Roasted peanutty		46.9	35.7	36.7
Raw/beany		19.4	25.6	22.4
Burnt		35.4	24.8	40.8

[1] Intensity scores are based on a 150-mm unstructured line scale.

[2] Factors are X_1 = oil level (%), X_2 = degree of roast (color lightness, L), X_3 = percent salt. Prediction models are: brown color = $10.16 + 6.32\ X_1 + 0.33\ X_2 - 0.074\ X_1X_2 - 0.042\ X_1X_1$; Roasted peanutty = $30.18 + 0.11\ X_1 + 0.64\ X_2 - 0.037\ X_1X_2 + 0.031\ X_1X_1$; Raw/beany = $43.96 - 1.78\ X_1 - 0.22\ X_2 + 0.019\ X_1X_2 + 0.015\ X_1X_1$; Burnt = $330.15 - 0.56\ X_1 - 9.80\ X_2 - 4.086\ X_3 + 0.08\ X_2X_3 + 0.083\ X_2X_2$.

[3] A = minimum oil and dark degree of roast for peanuts with a hedonic rating of 6 (like slightly) on a 9-point scale for overall acceptance; B = minimum oil and light degree of roast for an acceptable product; C = oil level and degree of roast for an acceptable product with 1/3 less oil than a full fat, roasted, salted peanut at 48% oil.

attribute for degree of roast has a negative value as expected, considering that a higher color lightness measurement indicates a lower degree or roast; higher levels or roast are important factors in increased intensity of the burnt flavor. The parameter estimate for raw/beany and burnt attributes for degree of roast and level of oil have negative values, indicating the attribute intensities will, respectively, decrease with higher oil levels and a higher (lighter) degree of roast. The interaction term between oil level and degree of roast is not a highly important factor affecting descriptive scores.

Scores equivalent to a maximum of 6 for overall liking, where maximum and minimum acceptable oil levels were 39.1% with a dark degree of roast (L = 44) and 29.2% with a light degree or roast (L = 56), were substituted into the regression models for brown color, roasted peanutty, raw/beany, and burnt attributes. Attribute scores from descriptive analysis prediction equations for an acceptable product are presented in Table 13.3. Sample A represents a product with the minimum oil at the darkest roast level (L = 44) for an acceptable quality roasted, defatted, salted peanut. Sample B represents a product with the minimum oil level at the lightest roast level (L = 56) for an acceptable quality product, and sample C represents a reduced oil level peanut with one-third less oil. To obtain an acceptable product (overall acceptance = 6), peanuts with 32% oil must have a degree or roast resulting in a color lightness, L, no less than 54.

Factor levels for products A and B were substituted into the prediction equations for brown color, roasted peanutty, raw/beany, and burnt attributes. Results indicated that these defatted, roasted, salted peanuts had a brown color not too light or dark (descriptive score = 56.8–81.0), a roasted peanutty intensity of 35.7–46.9, and a raw/beany and burnt intensity that is only slightly detectable (descriptive score = 19.4–25.6, and 24.8–35.4, respectively). The attribute intensity scores for sample C falls within the range of descriptive scores from samples A and B.

Attributes of acceptable defatted, roasted, salted peanuts were quantified through prediction equations. RSM allows the researcher to quantify the effect of multiple factors important in formulating a quality roasted, defatted, salted peanut simultaneously and in considerably less time than required to graph acceptance scores versus descriptive attribute score individually.

Conclusion

In the development of high-quality food products, a strategic approach that involves the identification of critical attributes is recommended. Consumer acceptance of the food or food product can be quantified by using consumer affective tests and characterizing this using sensory descriptive analysis ratings or instrumental and physicochemical measurements. The relationships can be determined by plotting acceptance ratings against the descriptive analysis ratings or the physicochemical measurements. Mathematical models may be developed that can be used to predict consumer acceptance scores from descriptive analysis ratings or physicochemical measurements. These equations may be used to establish specifications for food that correspond to a predetermined degree of quality.

References

ASTM. 1979. ASTM Manual on Consumer Sensory Evaluation, ASTM Special Technical Publication 682, E. E. Schaefer, ed. American Society for Testing and Materials, Philadelphia, PA. pp. 28–30.

ASTM. 1993. Manual on Descriptive Analysis Testing for Sensory Evaluation. ASTM Committee E-18 on Sensory Evaluation. American Society for Testing and Materials, Philadelphia, PA.

Bastos, D. H. M., Domenech, C. H., and Areas, J. A. G. 1991. Optimization of extrusion cooking of lung proteins by response surface methodology. *Intl. J. Food Sci. Technol.* 26:403–408.

Bhattacharya, M., Hanna, M. A., and Kaufman, R. E. 1986. Textural properties of extruded plant protein blends. *J. Food Sci.* 51:988–993.

Bourne, M. C. 1982. *Food Texture and Viscosity.* Academic Press, New York.

Brandt, M. A., Skinner, E. Z., and Coleman, J. A. 1963. Texture profile method. *J. Food Sci.* 28:404–409.

Bruns, A. J., and Bourne, M. C. 1975. Effects of sample dimensions on the snapping force of crisp foods. *J. Texture Stud.* 68:445–458.

Cairncross, S. E., and Sjostrom, L. B. 1950. Flavor profiles—a new approach to flavor problems. *Food Technol.* 4:308–311.

Cardello, A. V., Maller, O., Kapsalis, J. G., Segars, R. A., Sawyer, F. M., Murphy, C., and Moskowitz, H. R. 1982. Perception of texture by trained and consumer panelists. *J. Food Sci.* 47:1186–1197.

Caul, J. F. 1957. The profile method in flavor analysis. *Adv. Food Res.* 7:1–6.

Chen, A. W., Resurreccion, A. V. A., and Paguio, L. P. 1996. Age appropriate hedonic scales to measure food preferences of young children. *J. Sens. Stud.* 11:141–163.

Chow, E. T. S., Wei, L. S., Devor, R. E., and Steinberg, M. P. 1988. Performance of ingredients in a soybean whipped topping: A response surface analysis. *J. Food Sci.* 53:1761–1765.

Christensen, C. M. 1983. Effects of color on aroma, flavor and texture judgments of foods. *J. Food Sci.* 48:787–790.

Clydesdale, F. M. 1978. Colorimetry-methodology and applications. *Crit. Rev. Food Sci. Nutr.* 10:243–301.

Clydesdale, F. M. 1991. Color perception and food quality. *J. Food Qual.* 14:61–74.

Clydesdale, F. M. 1993. Color as a factor in food choice. *Crit. Rev. Food Sci. Nutr.* 33(1):83–101.

Curley, L. P., and Hoseney, R. C. 1984. Effects of corn sweeteners on cookie quality. *Cereal Chem.* 61(4):274–278.

Duxbury, D. D. 1988. R&D directions for the 1990's. *Food Processing* 49(8):19–28.

Floros, J. D., and Chinnan, M. S. 1988. Seven factor response surface optimization of a double-stage lye (NaOH) peeling process for pimiento peppers. *J. Food Sci.* 53:631–638.

Freund, R. J., and Littel, R. C. 1986. *System for Regression.* SAS Institute, Cary, NC.

Frijters, J. E. R. 1988. A review of Roland Harper's research in psychology and food science. In *Food Acceptability.* D. M. H. Thomson, ed., Elsevier Science Publishers, New York. pp. 11–25.

Galvez, F. C. F., and Resurreccion, A. V. A. 1992. Reliability of the focus group technique in determining the quality characteristics of mungbean (*Vigna radiata* (L). Wilczec) noodles. *J. Sens. Stud.* 7:315–326.

Galvez, F. C. F., Resurreccion, A. V. A., and Koehler, P. E. 1990. Optimization of processing of peanut beverage. *J. Sens. Stud.* 5:1–17.

Gormley, T. R. 1987. Fracture testing of cream cracker biscuits. *J. Food Eng.* 6:325–332.

Harper, R. 1981. The nature and importance of individual differences. In *Criteria of Food Acceptance: How Man Chooses What He Eats,* J. Solms and R. L. Hall, eds. Forster Publishing, Zurich, pp. 220–237.

Henselman, M. R., Donatoni, S. M., and Henika, R. G. 1974. Use of response surface methodology in the development of acceptable high protein bread. *J. Food Sci.* 39:943–946.

Hill, W. J., and Hunter, W. G. 1966. A review of response surface methodology: A literature survey. *Technometrics* 8(4):571–590.

IFT. 1990. Quality of fruits and vegetables: A scientific status summary by the Institute of Food Technologists Expert Panel on Food Safety and Nutrition. N. H. Mermelstein, ed. *Food Technol.* 44(6):99–106.

IFT/SED. 1981. Guidelines for the preparation and review of papers reporting sensory evaluation data. *Food Technol.* 35(11):50–59.

Johnson, T. M., and Zabik, M. E. 1981. Response surface methodology for analysis of protein interactions in angel food cakes. *J. Food Sci.* 46:1226–1230.

Katz, E. E., and Labuza, T. P. 1981. Effect of water activity on the sensory crispness and mechanical deformation of snack food products. *J. Food Sci.* 46:403–409.

Kimmel, S. A., Sigman-Grant, M., and Guinard, J. X. 1994. Sensory testing with young children. *Food Technol.* 48(3):92–99.

Kroll, B. J. 1990. Evaluating rating scales for sensory testing with children. *Food Technol.* 44(11):78–86.

Malundo, T. M. M., and Resurreccion, A. V. A. 1992. A comparison of performance of panels selected using analysis of variance and cluster analysis. *J. Sens. Stud.* 7:63–65.

Meilgaard, M., Civille, G. V., and Carr, B. T. 1987. Sensory Evaluation Techniques, Vol. II. CRC Press, Boca Raton, FL. pp. 1–24.

Meilgaard, M., Civille, G. V., and Carr, B. T. 1991. *Sensory Evaluation Techniques,* 2nd ed. CRC Press, Boca Raton, FL.

Moskowitz, H. 1974. Sensory evaluation by magnitude estimation. *Food Technol.* 28(11):16, 18, 20–21.

Mudahar, G. S., Toledo, R. T., and Jen, J. J. 1990. A response surface methodology approach to optimize potato dehydration process. *J. Food Process. Preserv.* 14:93–106.

Munoz, A. M., and Chambers, E., IV. 1993. Relating sensory measurements to consumer acceptance of meat products. *Food Technol.* 47(11):128–131, 134.

Munoz, A. J., Civille, G. V., and Carr, B. T. 1992. *Sensory Evaluation in Quality Control.* Van Nostrand Reinhold, New York.

Oh, N. H., Seib, P. A., and Chung, D. S. 1985. Noodles: III. Effects of processing variables on quality characteristics of dry noodles. *Cereal Chem.* 62:437–440.

O'Mahony, M. 1991. Descriptive analysis and concept alignment. In *IFT Basic Symposium Series: Sensory Science Theory and Applications in Foods,* H. T. Lawless and B. P. Klein, eds. Marcel Dekker, New York, Basel, and Hong Kong, pp. 223–267.

Plemmons, L. P. 1997. Sensory evaluation methods to improve validity, reliability, and interpretation of panelist responses. M.S. thesis, University of Georgia, Athens, GA.

Resurreccion, A. V. A. 1997. Sensory evaluation methods to measure quality of frozen food. In *Quality in Frozen Food,* M.C. Erickson and Y.-C. Hung, eds. Chapman & Hall, New York.

SAS Institute. 1980. *SAS Applications,* SAS Institute, Cary, NC.

Schutz, H. G. 1983. Multiple regression approach to optimization. *Food Technol.* 37(11):46–48, 62.

Seymour, S. K., and Hamann, D. D. 1988. Crispness and crunchiness of selected low moisture foods. *J. Texture Stud.* 19:79–95.

Shelke, K., Dick, J. W., Holm, Y. F., and Loo, K. S. 1990. Chinese wet noodle formulation: A response surface methodology study. *Cereal Chem.* 67:338–342.

Sherman, P., and Deghaidy, F. S. 1978. Force-deformation conditions associated with the evaluation of brittleness and crispness in selected foods. *J. Texture Stud.* 9:437–459.

Sokolow, H. 1988. Qualitative methods for language development. In *Applied Sensory Analysis of Foods,* Vol. I., H. Moskowitz, ed. CRC Press, Boca Raton, FL. pp. 4–20.

Stone, H. 1992. Quantitative descriptive analysis (QDA). In *Manual on Descriptive Analysis Testing for Sensory Evaluation,* R. C. Hootman, ed. ASTM, Philadelphia, PA.

Stone, H., and Sidel, J. L. 1993. Affective testing. In *Sensory Evaluation Practices,* 2nd ed. Academic Press, New York, NY.

Stone, H., Sidel, J., Oliver, S., Woolsey, A., and Singleton, R. C. 1974. Sensory evaluation by quantitative descriptive analysis. *Food Technol.* 28(11):24–34.

Szczesniak, A. S., Brandt, M. A., and Friedman, H. H. 1963. Development of standard rating scales for mechanical parameters of texture and correlation between the objective and the sensory methods of texture evaluation. *J. Food Sci.* 28:397–403.

Vainionpaa, J. 1991. Modelling of extrusion cooking of cereals using response surface methodology. *J. Food Eng.* 13:1–26.

Vaisey-Genser, M., Ylimaki, G., and Johnston, B. 1987. The selection of levels of canola oil, water, and an emulsifier system in cake formulations by response surface methodology. *Cereal Chem.* 64:50–54.

Ward, C. D. W. 1995. Effect of physicochemical and sensory properties on functionality and consumer acceptance of snack chips formulated with cowpea, corn, and wheat flour mixtures. Ph.D. dissertation, University of Georgia, Athens, GA.

Appendix A
Checklist for
Focus Group Study

PLANNING

DATE COMPLETED	TASK TO BE PERFORMED
_____	Start a project journal, maintain a project file or file box for tapes and all project materials.
_____	Verify approval status for research with human subjects.
_____	Hold a planning meeting with client or requester; verify objectives.
_____	Determine critical questions to be asked; samples to be served, if any; finalize recruitment criteria for test.
_____	Prepare test protocol outline.
_____	Schedule sensory test dates (consider school and community events).
_____	Schedule facilities, check focus group room, kitchen facilities as needed.
_____	Finalize personnel needs.
_____	Assign jobs to project personnel, prepare duty roster with activity and name of person involved.
_____	Schedule tasks, include dry-run date(s).

RECRUITMENT

DATE COMPLETED	TASK TO BE PERFORMED
_____	Prepare a Recruitment Screener.
_____	Recruit consumers by telephone using database; make any corrections on consumer database if needed; initial your name and date your update to database.

DEVELOPMENT OF QUESTIONNAIRE AND MODERATORS GUIDE

DATE COMPLETED   **TASK TO BE PERFORMED**

_____   Develop scannable demographic questionnaire.

_____   Develop moderators' guide; obtain approval from requester or client.

TEST SUPPLIES AND PANELIST INCENTIVES

DATE COMPLETED   **TASK TO BE PERFORMED**

_____   Coordinate with budget office for panelist incentives.

_____   Request for cash for panelist incentives, amount of money needed, dates needed and denominations of bills for panelists.

_____   Prepare a supply and equipment list (tape recorders, samples, preparation and serving supplies, office supplies, tapes, etc.).

_____   Obtain approvals for purchase of items on supply list; write purchase orders.

_____   Purchase supplies; keep copies of documentation in project file.

TEST PREPARATION

DATE COMPLETED   **TASK TO BE PERFORMED**

_____   Finalize detailed project outline: specify in detail, project title; objectives of project; each activity involved, such as sample preparation, testing procedures, setup and dry-run, work schedule for each activity (starting time and time to be completed); assignments; work request; test schedules, etc.

_____   Generate and assign three-digit random numbers for samples as needed.

_____   Label serving dishes, as needed.

_____   Prepare written instructions and scripts to be distributed to project personnel.

_____   Prepare master copy of receipt forms for incentives.

DATE COMPLETED   **TASK TO BE PERFORMED**

_____   Duplicate necessary documents:
Demographic questionnaire, ivory (5 pages per panelist)
Scoresheet(s), white (vary)
Receipt forms for incentives (1 copy per panelist)
Consent forms, yellow (2 copies per panelist)

_____   Label scannable scoresheets with sample number.

_____   Prepare large-print name tags.

_____   Visit central location site, set up, and conduct a complete briefing and dry-run one to three days before test date.

_____   Complete all remaining tasks.

DAY AFTER TEST

DATE COMPLETED   **TASK TO BE PERFORMED**

_____   Scan demographic questionnaires.

_____   Organize all files and materials for project file box.

DATA ANALYSIS AND REPORT PREPARATION

DATE COMPLETED   **TASK TO BE PERFORMED**

_____   Transcribe tapes.

_____   Print out all results; place a hard copy in project file box and save a copy to a disk.

_____   Prepare report.

Appendix B
Checklist for Consumer Laboratory Test

PLANNING

DATE COMPLETED	TASK TO BE PERFORMED
_____	Start a project file, maintain a project file box for questionnaires and all project materials.
_____	Verify approval status for research with human subjects; keep a copy in the project file.
_____	Determine (to be approved by client) sample and serving conditions, weight or volume to be served, temperature, maximum holding time, special conditions.
_____	Hold a planning meeting with client or requester. Verify objectives. Finalize experimental design.
_____	Finalize recruitment criteria and panelist quotas for test.
_____	Finalize test protocol in outline form: Determine sample preparations and serving conditions; weight or volume to be served, temperature, maximum holding time, special serving conditions.
_____	Schedule sensory test dates on calendar (consider school and community events).
_____	Schedule facilities: Reserve reception room, kitchens, and booths.
_____	Finalize personnel needs.
_____	Assign jobs to project personnel: Prepare duty roster with activity and name of person involved.
_____	Schedule tasks for test: Include dry-run date(s).

RECRUITMENT

DATE COMPLETED   **TASK TO BE PERFORMED**

_____   Prepare a recruitment screener.

_____   Recruit consumers by telephone using database; make corrections on consumer database if needed. Initial your name and date your update to database.

EXPERIMENTAL DESIGN AND QUESTIONNAIRE DEVELOPMENT

DATE COMPLETED   **TASK TO BE PERFORMED**

_____   Develop sample presentation design; randomization or counterbalancing scheme for samples, treatments, blocking. Generate 3-digit sample codes.

_____   Design scannable or computerized scoresheets; pretest scoresheets.

_____   Develop scannable demographic questionnaire.

TEST SUPPLIES AND PANELIST INCENTIVES

DATE COMPLETED   **TASK TO BE PERFORMED**

_____   Coordinate with budget office for panelist incentives; request for petty cash for panelist incentives; advising amount of money needed, including date(s) and denomination ($5, $10, $20) you prefer to give panelists.

_____   Prepare a equipment and supply lists (specialized equipment, serving food preparation, and office supplies).

_____   Obtain approval for purchase of items on supply; write purchase order.

_____   Purchase supplies; keep documentation for reimbursement, placements in project file.

TEST PREPARATION

DATE COMPLETED   **TASK TO BE PERFORMED**

_____   Finalize detailed project outline (specify in detail, such as project title, objectives of project: each activity involved, such as sample preparation, testing procedures, setup, and dry run; work schedule for each activity, starting time

DATE COMPLETED	TASK TO BE PERFORMED
	and time to be finished; job assignments; work request; test schedules, etc.).
_____	Label survey containers.
_____	Prepare written instructions and scripts to be distributed to project personnel.
_____	Conduct dry-run of food preparation methods.
_____	Prepare cooking and sensory evaluation flow diagrams to be posted in appropriate areas.
_____	Conduct sample preparation dry-run.
_____	Finalize sign-in sheet for each test session and give copy to greeter.
_____	Prepare reminder cards or letters for consumers.
_____	Address and mail consumer reminders.
_____	Prepare panelist informed consent form.
_____	Prepare instruction sheet for consumer forms.
_____	Organize and duplicate all documents for duplication. Consent form: yellow, 2 copies per panelist. Receipt forms for incentives: 1 copy per panelist. Demographic questionnaire: ivory, 5 pages per panelist. Scoresheet: white, as needed
_____	Label consumer folders.
_____	Label scannable questionnaires with panelist ID and scoresheets with panelist ID and sample numbers.
_____	Assemble consumer folders.
_____	Obtain samples for test and store under appropriate conditions.
_____	Call consumers to remind them about their test schedules and read all information shown on the reminder card to panelist (optional).
_____	Set up and conduct a complete briefing and dry-run of all project personnel no less than three days before test is conducted.
_____	Prepare directional signs for consumers. This is especially important when two or more tests are scheduled on the same day.
_____	Set up reception room and sensory booth areas.
_____	Post serving scheme on server side of evaluation facilities.

DAY AFTER TEST

DATE COMPLETED	TASK TO BE PERFORMED
_____	Scan demographic questionnaires/scoresheets and/or collect data from computers.
_____	Organize all files and materials in project file box and glue any remaining loose information to project journal.

DATA ANALYSIS AND REPORT PREPARATION

DATE COMPLETED	TASK TO BE PERFORMED
_____	Prepare data codebook; use identical file name as corresponding data file but with "CDBK file type."
_____	Print out all raw data and data codebook. Place hard copy and diskette in project file box.
_____	Verify statistical analysis procedures with appropriate personnel.
_____	Conduct statistical analysis.
_____	Prepare figures and tables, and final report.
_____	Submit final report.

Appendix C
Checklist for Consumer Central Location Test (CLT)

PLANNING

DATE COMPLETED   **TASK TO BE PERFORMED**

_____   Start a project journal, maintain a project file box for questionnaires and all project materials.

_____   Verify approval status for research with human subjects; keep a copy in the project file box.

_____   Hold a planning meeting with all client or requester; verify objectives; finalize experimental design.

_____   Finalize recruitment criteria and panelist quotas for test.

_____   Finalize protocol outline; determine sample preparation and serving conditions, weight, volume to be served, temperature, maximum holding time, special serving conditions.

_____   Schedule test date(s), consider community and school events.

_____   Schedule CLT facility.

_____   Finalize personnel needs; assign jobs to project personnel, prepare duty; roster with activity and name of person involved.

_____   Schedule tasks, include dry run.

RECRUITMENT

DATE COMPLETED   **TASK TO BE PERFORMED**

_____   Prepare a recruitment screener.

EXPERIMENTAL DESIGN AND QUESTIONNAIRE DEVELOPMENT

<u>DATE COMPLETED</u>   <u>TASK TO BE PERFORMED</u>

_____   Develop experimental design.

_____   Develop scannable demographic questionnaire.

_____   Develop scannable scoresheet(s).

TEST SUPPLIES AND PANELIST INCENTIVES

<u>DATE COMPLETED</u>   <u>TASK TO BE PERFORMED</u>

_____   Coordinate with budget office for payment of incentives to consumers.

_____   Request for cash incentives for panelist amount of money needed, including date(s) needed and denominations ($5, $10 or $20) you prefer to give panelists.

_____   Prepare an equipment and supply list for food preparation, serving, and office supplies.

_____   Obtain approval for purchase of items on supply list.

_____   Write purchase orders and purchase supplies; keep documentation of expenses in project file.

TEST PREPARATION

<u>DATE COMPLETED</u>   <u>TASK TO BE PERFORMED</u>

_____   Finalize detailed project outline (specify in detail, project title, objectives of project, each activity involved, sample preparation, testing procedures, setup and dry run); work schedule for each task involved (starting time and time to be finished); assignments, work requests, test schedules, etc.

_____   Prepare sample serving scheme, randomization and counterbalancing scheme for samples, treatments, blocking. Generate and assign three-digit random numbers for sample codes.

_____   Label serving dishes.

_____   Prepare written instructions and scripts to be distributed to project personnel.

_____   Prepare sensory evaluation flow diagrams to be posted in appropriate areas.

DATE COMPLETED	TASK TO BE PERFORMED
_____	Prepare master copy of receipt forms for incentives.
_____	Duplicate necessary documents. Demographic questionnaire, ivory (5 pages per panelist) Scoresheet(s), white (vary) Receipt forms for incentives (1 copy per panelist)
_____	Label scannable scoresheets with sample numbers.
_____	Visit central location site, set up and conduct briefing and a complete dry-run 1–3 days before test date.
_____	Complete all remaining tasks.
_____	Post serving scheme on server side of evaluation facilities.

DAY AFTER TEST

DATE COMPLETED	TASK TO BE PERFORMED
_____	Scan demographic questionnaire and scoresheets.
_____	Organize all files and materials for project file box.

DATA ANALYSIS AND REPORT PREPARATION

DATE COMPLETED	TASK TO BE PERFORMED
_____	Prepare data codebook; use identical filename as corresponding data file but with "CDBK" file type.
_____	Print out all raw data and data codebook; place a hard copy in project file box and save a copy to a disk.
_____	Finalize plan for statistical analysis with appropriate personnel.
_____	Conduct statistical analysis.
_____	Submit final report.

Appendix D
Checklist for Consumer Home-Use Test (HUT)

PLANNING

DATE COMPLETED	TASK TO BE PERFORMED
_____	Start a project journal; maintain a project file or file box for questionnaires and all project materials.
_____	Verify approval status for research with human subjects; keep a copy in the project file.
_____	Hold a planning meeting with client or requester.
_____	Finalize recruitment criteria and sample quotas for test; verify objectives; finalize experimental design.
_____	Finalize test protocol outline and data-collection strategy.
_____	Determine samples, sample size, packaging material, and materials to go in package for respondent; decide on test period.
_____	Schedule test date, facilities as needed for distribution of samples.
_____	Finalize personnel needs; assign jobs to project personnel involved; prepare duty roster with activity and name of person assigned.
_____	Schedule tasks.

RECRUITMENT

DATE COMPLETED	TASK TO BE PERFORMED
_____	Prepare a recruitment screener.
_____	Prepare consumer recruitment calling sheets incl. consumer ID #s, names, addresses, telephone numbers, date(s) contacted, status of call(s), results (yes, no or alternate) and schedule.

<u>**DATE COMPLETED**</u>   <u>**TASK TO BE PERFORMED**</u>

——————————— Recruit consumers by telephone using database; make any corrections on consumer database as needed; initial your name and date your update in database.

QUESTIONNAIRE DEVELOPMENT

<u>**DATE COMPLETED**</u>   <u>**TASK TO BE PERFORMED**</u>

——————————— Develop scannable demographic questionnaire.

——————————— Train interviewers for data collection as needed.

TEST SUPPLIES AND PANELIST INCENTIVES

<u>**DATE COMPLETED**</u>   <u>**TASK TO BE PERFORMED**</u>

——————————— Prepare a equipment and supply list (mailers, sample containers, office supplies).

——————————— Obtain approval for purchase of items on supply list.

——————————— Write purchase orders.

——————————— Purchase supplies; keep documentation for reimbursement, place copies in project file.

TEST PREPARATION

<u>**DATE COMPLETED**</u>   <u>**TASK TO BE PERFORMED**</u>

——————————— Finalize project outline. Specify in detail: project title; objectives of project; each activity involved, such as sample preparation, testing procedures, setup and dry run; work schedule for each activity involved (starting time and time to be finished); assignments; work requests; test schedules, etc.

——————————— Generate and assign 3-digit random numbers for samples; label samples.

——————————— Prepare detailed instructions for panelists.

——————————— Prepare master copy of panelist instruction sheet.

——————————— Organize all documents for duplication.
Documents to be duplicated:
Panelist instruction sheet, white (1 copy per panelist)
Demographic questionnaire, Ivory (5 pages per panelist)
Scoresheet(s): varies with test.

DATE COMPLETED	TASK TO BE PERFORMED
_____	Duplicate all necessary documents.
_____	Label scannable questionnaires/scoresheets with sample # and/or panelist ID #.
_____	Contribute sample packages.
_____	Call consumers to remind them to return their questionnaire scoresheets come back to pick up sample for a second test.

DAY AFTER TEST

DATE COMPLETED	TASK TO BE PERFORMED
_____	Receive all questionnaires.
_____	Scan questionnaire/scoresheets.
_____	Organize all files and materials for project file box.
_____	Glue all remaining loose information to project journal.

DATA ANALYSIS AND REPORT PREPARATION

DATE COMPLETED	TASK TO BE PERFORMED
_____	Prepare data codebook. Use identical file name as corresponding data file but with "CDBK" file type.
_____	Print out all raw data and data codebook.
_____	Finalize plan for statistical analysis with appropriate personnel.
_____	Conduct statistical analysis.
_____	Prepare tables, figures and final report.
_____	Submit final report.

Appendix E
Statistical Tables

Table E.1 Minimum Numbers of Agreeing Judgments Necessary to Establish Significance at Various Probability Levels for the Paired-Preference Test (Two-Tailed, $p = .05$)[a]

Number of Trials (n)	Probability Levels		Number of Trials (n)	Probability Levels	
	.05	.01		.05	.01
7	7		32	23	24
8	8	8	33	23	25
9	8	9	34	24	25
10	9	10	35	24	26
11	10	11	36	25	27
12	10	11	37	25	27
13	11	12	38	26	28
14	12	13	39	27	28
15	12	13	40	27	29
16	13	14	41	28	30
17	13	15	42	28	30
18	14	15	43	29	31
19	15	16	44	29	31
20	15	17	45	30	32
21	16	17	46	31	33
22	17	18	47	31	33
23	17	19	48	32	34
24	18	19	49	32	34
25	18	20	50	33	35
26	19	20	60	39	41
27	20	21	70	44	47
28	20	22	80	50	52
29	21	22	90	55	58
30	21	23	100	61	64
31	22	24			

[a]Value (x) not appearing in table may be derived from $x = [Z[(n + n + 1)/2]^{1/2}]$, where n = number of trials, x = minimum number of correct judgments, if $x = 1.96$ at a probability (α) = 5%, and $Z = 2.58$ at probability (α) = 1%. Adapted with permission from Rosseler et al. (1978).

242 _equation is rubbish_ ?

where in the reference?

Table E.2 Critical Values of Differences Between Rank Sums, $p = .05$[a]

No. of Assessors	No. of Products								
	2	3	4	5	6	7	8	9	10
20	8.8	14.8	21.0	27.3	33.7	40.3	47.0	53.7	60.6
21	9.0	15.2	21.5	28.0	34.6	41.3	48.1	55.1	62.1
22	9.2	15.5	22.0	28.6	35.4	42.3	49.2	56.4	63.5
23	9.4	15.9	22.5	29.3	36.2	43.2	50.3	57.6	65.0
24	9.6	16.2	23.0	29.9	36.9	44.1	51.4	58.9	66.4
25	9.8	16.6	23.5	30.5	37.7	45.0	52.5	60.1	67.7
26	10.0	16.9	23.9	31.1	38.4	45.9	53.5	61.3	69.1
27	10.2	17.2	24.4	31.7	39.2	46.8	54.6	62.4	70.4
28	10.4	17.5	24.8	32.3	39.9	47.7	55.6	63.6	71.7
29	10.6	17.8	25.3	32.8	40.6	48.5	56.5	64.7	72.9
30	10.7	18.2	25.7	33.4	41.3	49.3	57.5	65.8	74.2
31	10.9	18.5	26.1	34.0	42.0	50.2	58.5	66.9	75.4
32	11.1	18.7	26.5	34.5	42.6	51.0	59.4	68.0	76.6
33	11.3	19.0	26.9	35.0	43.3	51.7	60.3	69.0	77.8
34	11.4	19.3	27.3	35.6	44.0	52.5	61.2	70.1	79.0
35	11.6	19.6	27.7	36.1	44.6	53.3	62.1	71.1	80.1
36	11.8	19.9	28.1	36.6	45.2	54.0	63.0	72.1	81.3
37	11.9	20.2	28.5	37.1	45.9	54.8	63.9	73.1	82.4
38	12.1	20.4	28.9	37.6	46.5	55.5	64.7	74.1	83.5
39	12.2	20.7	29.3	38.1	47.1	56.3	65.6	75.0	84.6
40	12.4	21.0	29.7	38.6	47.7	57.0	66.4	76.0	85.7
41	12.6	21.2	30.0	39.1	48.3	57.7	67.2	76.9	86.7
42	12.7	21.5	30.4	39.5	48.9	58.4	68.0	77.9	87.8
43	12.9	21.7	30.8	40.0	49.4	59.1	68.8	78.8	88.8
44	13.0	22.0	31.1	40.5	50.0	59.8	69.6	79.7	89.9
45	13.1	22.2	31.5	40.9	50.6	60.4	70.4	80.6	90.9
46	13.3	22.5	31.8	41.4	51.1	61.1	71.2	81.5	91.9
47	13.4	22.7	32.2	41.8	51.7	61.8	72.0	82.4	92.9
48	13.6	23.0	32.5	42.3	52.2	62.4	72.7	83.2	93.8
49	13.7	23.2	32.8	42.7	52.8	63.1	73.5	84.1	94.8
50	13.9	23.4	33.2	43.1	53.3	63.7	74.2	85.0	95.8
55	14.5	24.6	34.8	45.2	55.9	66.8	77.9	89.1	100.5
60	15.2	25.7	36.3	47.3	58.4	69.8	81.3	93.1	104.9
65	15.8	26.7	37.8	49.2	60.8	72.6	84.6	96.9	109.2
70	16.4	27.7	39.2	51.0	63.1	75.4	87.8	100.5	113.3
80	17.5	29.6	42.0	54.6	67.4	80.6	93.9	107.5	121.2
90	18.6	31.4	44.5	57.9	71.5	85.5	99.6	114.0	128.5
100	19.6	33.1	46.9	61.0	75.4	90.1	105.0	120.1	135.5
110	20.6	34.8	49.2	64.0	79.1	94.5	110.1	126.0	142.1
120	21.5	36.3	51.4	66.8	82.6	98.7	115.0	131.6	148.4

[a]From D. Basker. 1988. Critical values of differences among rank sums for multiple comparisons, *Food Tech.* 42(2):79, Table 1. Reprinted with permission of the Institute of Food Technologists.

Table E.3 Percentage Points of the F-Distribution[a]

v^2	1	2	3	4	5	10	20
				Upper 5% points			
5	6.61	5.79	5.41	5.19	5.05	4.74	4.56
6	5.99	5.14	4.76	4.53	4.39	4.06	3.87
7	5.59	4.74	4.35	4.12	3.97	3.64	3.44
8	5.32	4.46	4.07	3.84	3.69	3.35	3.15
9	5.12	4.26	3.86	3.63	3.48	3.14	2.94
10	4.96	4.10	3.71	3.48	3.33	2.98	2.77
11	4.84	3.98	3.59	3.36	3.20	2.85	2.65
12	4.75	3.89	3.49	3.26	3.11	2.75	2.54
13	4.67	3.81	3.41	3.18	3.03	2.67	2.46
14	4.60	3.74	3.34	3.11	2.96	2.60	2.39
15	4.54	3.68	3.29	3.06	2.90	2.54	2.33
16	4.49	3.63	3.24	3.01	2.85	2.49	2.28
17	4.45	3.59	3.20	2.96	2.81	2.45	2.23
18	4.41	3.55	3.16	2.93	2.77	2.41	2.19
19	4.38	3.52	3.13	2.90	2.74	2.38	2.16
20	4.35	3.49	3.10	2.87	2.71	2.35	2.12
21	4.32	3.47	3.07	2.84	2.68	2.32	2.10
22	4.30	3.44	3.05	2.82	2.66	2.30	2.07
23	4.28	3.42	3.03	2.80	2.64	2.27	2.05
24	4.26	3.40	3.01	2.78	2.62	2.25	2.03
25	4.24	3.39	2.99	2.76	2.60	2.24	2.01
26	4.23	3.37	2.98	2.74	2.59	2.22	1.99
27	4.21	3.35	2.96	2.73	2.57	2.20	1.97
28	4.20	3.34	2.95	2.71	2.56	2.19	1.96
29	4.18	3.33	2.93	2.70	2.55	2.18	1.94
30	4.17	3.32	2.92	2.69	2.53	2.16	1.93
40	4.08	3.23	2.84	2.61	2.45	2.08	1.84
60	4.00	3.15	2.76	2.53	2.37	1.99	1.75
120	3.92	3.07	2.68	2.45	2.29	1.91	1.66
∞	3.84	3.00	2.60	2.37	2.21	1.83	1.57

(continued)

Table E.3   (*Continued*)

v^2	1	2	3	4	5	10	20
				Upper 1% points			
5	16.26	13.27	12.06	11.39	10.97	10.05	9.55
6	13.75	10.92	9.78	9.15	8.75	7.87	7.40
7	12.25	9.55	8.45	7.85	7.46	6.62	6.16
8	11.26	8.65	7.59	7.01	6.63	5.81	5.36
9	10.56	8.02	6.99	6.42	6.06	5.26	4.81
10	10.04	7.56	6.55	5.99	5.64	4.85	4.41
11	9.65	7.21	6.22	5.67	5.32	4.54	4.10
12	9.33	6.93	5.95	5.41	5.06	4.30	3.86
13	9.07	6.70	5.74	5.21	4.86	4.10	3.66
14	8.86	6.51	5.56	5.04	4.69	3.94	3.51
15	8.68	6.36	5.42	4.89	4.56	3.80	3.37
16	8.53	6.23	5.29	4.77	4.44	3.69	3.26
17	8.40	6.11	5.18	4.67	4.34	3.59	3.16
18	8.29	6.01	5.09	4.58	4.25	3.51	3.08
19	8.18	5.93	5.01	4.50	4.17	3.43	3.00
20	8.10	5.85	4.94	4.43	4.10	3.37	2.94
21	8.02	5.78	4.87	4.37	4.04	3.31	2.88
22	7.95	5.72	4.82	4.31	3.99	3.26	2.83
23	7.88	5.66	4.76	4.26	3.94	3.21	2.78
24	7.82	5.61	4.72	4.22	3.90	3.17	2.74
25	7.77	5.57	4.68	4.18	3.85	3.13	2.70
26	7.72	5.53	4.64	4.14	3.82	3.09	2.66
27	7.68	5.49	4.60	4.11	3.78	3.06	2.63
28	7.64	5.45	4.57	4.07	3.75	3.03	2.60
29	7.60	5.42	4.54	4.04	3.73	3.00	2.57
30	7.56	5.39	4.51	4.02	3.70	2.98	2.55
40	7.31	5.18	4.31	3.83	3.51	2.80	2.37
60	7.08	4.98	4.13	3.65	3.34	2.63	2.20
120	6.85	4.79	3.95	3.48	3.17	2.47	2.03
∞	6.63	4.61	3.78	3.32	3.02	2.32	1.88

[a]This table is abridged from Table 18 of the *Biometrika Tables for Statisticians,* Vol. 1, 3rd ed., edited by E. S. Pearson and H. O. Hartley. Reproduced by permission of the Biometrika Trustees.

Table E.4   Table of Critical Values of t[a]

	Level of Significance for One-Tailed Test			
	.10	.025	.01	.005
df	Level of Significance for Two-Tailed Test			
	.10	.05	.02	.01
1	6.314	12.706	31.821	63.657
2	2.920	4.303	6.965	9.925
3	2.353	3.182	4.541	5.841
4	2.132	2.776	3.747	4.604
5	2.015	2.571	3.365	4.032
6	1.943	2.447	3.143	3.707
7	1.895	2.365	2.998	3.499
8	1.860	2.306	2.896	3.355
9	1.833	2.262	2.821	3.250
10	1.812	2.228	2.764	3.169
11	1.796	2.201	2.718	3.106
12	1.782	2.179	2.681	3.055
13	1.771	2.160	2.650	3.012
14	1.761	2.145	2.624	2.977
15	1.753	2.131	2.602	2.947
16	1.746	2.120	2.583	2.921
17	1.740	2.110	2.567	2.898
18	1.734	2.10	2.552	2.878
19	1.729	2.093	2.539	2.861
20	1.725	2.086	2.528	2.845
21	1.721	2.080	2.518	2.831
22	1.717	2.074	2.508	2.819
23	1.714	2.069	2.500	2.807
24	1.711	2.064	2.492	2.797
25	1.708	2.060	2.485	2.787
26	1.706	2.056	2.479	2.779
27	1.703	2.052	2.473	2.771
28	1.701	2.048	2.467	2.763
29	1.699	2.045	2.462	2.756
30	1.697	2.042	2.457	2.750
40	1.684	2.021	2.423	2.704
60	1.671	2.000	2.390	2.660
120	1.658	1.980	2.358	2.617
∞	1.645	1.960	2.326	2.576

[a]Reprinted from E. S. Pearson and H. O. Hartley, *Biometrika, Tables for Statisticians,* Vol. 1, 3rd ed., 1966, with permission of the Trustees of Biometrika.

Table E.5 Table of Critical Values of Chi-Square[a]

df	Probability under H_0 that $x^2 \geq$ chi-square	
	.05	.01
1	3.84	6.64
2	5.99	9.21
3	7.82	11.34
4	9.49	13.28
5	11.07	15.09
6	12.59	16.81
7	14.07	18.48
8	15.51	20.09
9	16.92	21.67
10	18.31	23.21
11	19.68	24.72
12	21.03	26.22
13	22.36	27.69
14	23.68	29.14
15	25.00	30.58
16	26.30	32.00
17	27.59	33.41
18	28.87	34.80
19	30.14	36.19
20	31.41	37.57
21	32.67	38.93
22	33.92	40.29
23	35.17	41.64
24	36.42	42.98
25	37.65	44.31
26	38.88	45.64
27	40.11	46.96
28	41.34	48.28
29	42.56	49.59
30	43.77	50.89

Table E.6 Table of Critical Values of p, the Spearman Rank Correlation Coefficient, and r, the Pearson Product–Moment Correlation Coefficient[a]

N	P		r	
	.05	.01	.05	.01
4	1.000		.811	.917
5	.900	1.000	.754	.874
6	.829	.943	.707	.834
7	.714	.893	.666	.793
8	.643	.833	.632	.765
9	.600	.783	.602	.735
10	.564	.746	.576	.708
12	.506	.712	.532	.661
14	.456	.645	.497	.623
16	.425	.601	.468	.590
18	.399	.564	.444	.561
20	.377	.534	.423	.537
22	.359	.508	.404	.515
24	.343	.485	.388	.596
26	.329	.465	.374	.478
28	.317	.448	.361	.463
30	.306	.432	.349	.449

[a]From G. J. Glasser and R. F. Winter, *Biometrika,* Vol. 48. Reproduced by permission of the Biometrika Trustees.

INDEX